Olga Poleshchuk and Evgeniy Komarov

Expert Fuzzy Information Processing

T0205450

Studies in Fuzziness and Soft Computing, Volume 268

Editor-in-Chief

Prof. Janusz Kacprzyk
Systems Research Institute
Polish Academy of Sciences
ul. Newelska 6
01-447 Warsaw
Poland
E-mail: kacprzyk@ibspan.waw.pl

Further volumes of this series can be found on our homepage: springer.com

Olga Poleshchuk and Evgeniy Komarov

Expert Fuzzy Information Processing

 Springer

Authors

Prof. Olga Poleshchuk
Moscow State Forest University
1st Institutskaya st., 1
Mytishi
Moscow reg.
Russia
E-mail: olga.m.pol@yandex.ru

Dr. Evgeniy Komarov
Moscow State Forest University
1st Institutskaya st., 1
Mytishi
Moscow reg.
Russia
E-mail: komarov@mgul.ac.ru

ISBN 978-3-642-26787-1   ISBN 978-3-642-20125-7 (eBook)

DOI 10.1007/978-3-642-20125-7

Studies in Fuzziness and Soft Computing   ISSN 1434-9922

Typeset & Cover Design: Scientific Publishing Services Pvt. Ltd., Chennai, India.

Printed on acid-free paper

9 8 7 6 5 4 3 2 1

springer.com

Foreword

In essence, "Expert Fuzzy Information Processing," or FIP for short, a work co-authored by Professors Poleshchuk and Komarov, is an informative, authoritative, and up-to-date exposition of concepts and techniques which underlie processing of imperfect information—information which in one or more respects is uncertain, imprecise, fuzzy, unreliable or incomplete. In the real world, imperfect information is the rule rather than exception. Decisions are based in information. In this context, Professors Poleshchuk and Komarov's work is an important contribution to a better understanding of the remarkable human capability to make rational decisions in an environment of uncertainty and imprecision.

It is a deep-seated tradition in science to draw on probability theory—and only probability theory—to deal with problems in which uncertainty plays a significant role. The core of this tradition was eloquently stated by a prominent Bayesian, Professor Dennis Lindley: "The only satisfactory description of uncertainty is probability. By this I mean that every uncertainty statement must be in the form of a probability; that several uncertainties must be combined using the rules of probability; and that the calculus of probabilities is adequate to handle all situations involving uncertainty. Probability is the only sensible description of uncertainty and is adequate for all problems involving uncertainty. All other methods are inadequate...anything that can be done with fuzzy logic, belief functions, upper and lower probabilities, or any other alternative to probability can better be done with probability" (Lindley 1987). Is this a valid view? In the following, I take the liberty of arguing that the answer is No. In effect, No is the reason why the work of Professors Poleshchuk and Komarov is a significant addition to the armamentarium of standard probability theory.

The debut of fuzzy set theory in 1965 opened the door to development of fuzzy logic. Basically, fuzzy logic is the logic of classes with unsharp boundaries. Fuzziness—unsharpness of class boundaries—is distinct from randomness. A class, A, is precisiated (graduated) through association with degrees (grades) of membership. Humans have remarkable capability to graduate perceptions. If I am asked to mark on the scale from 0 to 1, the degree to which I like my job, I would have no difficulty in putting a crisp mark at, say, 0.7, or if allowed, a fuzzy mark centering on 0.7, with the understanding that fuzziness of the mark reflects my uncertainty about the degree to which I like my job. There is no randomness and no probability.

A concept which plays a pivotal role in fuzzy logic, but is absent in probability theory, is the concept of a linguistic variable. Informally, a linguistic variable is a variable whose values are words or phrases in a natural language. For example, Age is a linguistic variable if its values are young, old, middle-aged, very old, not very old, etc., with the understanding that such values are labels of fuzzy sets in the space of reals. When probability is treated as a linguistic variable, its values are likely, unlikely, not very likely, usually, etc. Linguistic probabilities are closely related to fuzzy quantifiers: most, many, few, not very many, about 5, etc. A simple example of a problem which involves linguistic probabilities and fuzzy quantifiers is the following. A box contains twenty black and white balls. What I know is that most of the balls are white. I draw a ball at random, note its color, and put it back. What is the probability that in thirty draws of balls most are black? The concept of a linguistic variable opens the door to computation with information described in a natural language. Example: Usually, Robert leaves office at about 5 pm. Usually, it takes Robert about an hour to get home from work. What is the probability that Robert is home at 6:15 pm? Many real world probabilities are perception-based and imprecise. The capability of fuzzy logic to compute with such probabilities is of major importance in the realm of decision-making in an environment of uncertainty and imprecision.

Another concept which plays an important role in fuzzy logic is that of possibility, with the understanding that possibility is a matter of degree. The concept of possibility is distinct from the concept of probability. In everyday discourse, we frequently use expressions such as "It is possible, but not probable that ...." In fuzzy logic we commonly have to deal with possibility distributions of probability distributions. Example. Consider the proposition: It is very likely that Robert is tall, where the fuzzy set tall plays the role of the possibility distribution of the variable Height(Robert). In this proposition, very likely plays the role of fuzzy probability of the fuzzy event Height(Robert) is tall. This statement may be interpreted as the possibility distribution of the probability distributions of the variable Height(Robert). The concept of a possibility distribution plays a key role in semantics of natural languages.

Indisputably, probability theory is a powerful tool for dealing with uncertainty and imprecision. But what the foregoing examples point to is that standard probability theory has serious limitations when it comes to dealing with problems in which uncertainty is caused by fuzziness, that is, unsharpness of class boundaries. To circumvent these limitations, it is necessary to add to the armamentarium of probability theory concepts and techniques drawn from fuzzy logic. In essence, this is what the important work of Professors Poleshchuk and Komarov serves to do.

FIP contains a wealth of information which is new and original. Particularly worthy of note is the concept of complete orthogonal semantic spaces which is introduced in Chapter 2 and is developed in Chapter 3. Another important concept is that of generalized models and rating systems of qualitative and quantitative characteristics of groups of objects. A topic which plays an important role in FIP is that of multiple hybrid fuzzy least-squares regression. Numerous examples are worked out in detail. An important feature of FIP is its skillful combination of formal theory and practical applications.

The work of Professors Poleshchuk and Komarov is not light reading, but the wealth of information which it presents is worth of careful study by readers who are faced with problems which standard probability theory cannot or does not address. Professors Poleshchuk and Komarov and the publisher, Springer, deserve our thanks and congratulations for producing a book which contributes so much to our ability to deal with problems in which the decision-relevant information is uncertain and/or imprecise. In the real world, such information is the rule rather than exception.

December 24, 2010   Lotfi A. Zadeh
                                                                             Berkeley, CA

Preface

This book deals with expert evaluation models in the form of semantic spaces with completeness and orthogonality properties (complete orthogonal semantic spaces). Theoretical and practical studies of some researchers have shown that these spaces describe expert evaluations most adequately, and as a result they were often included in more sophisticated models of intellectual systems for decision making and data analysis. Methods for constructing expert evaluation models of characteristics, comparative analysis of these models, studies of structural composition of their sets and constructing of generalized models are described. Models to obtain rating points for objects and groups of objects with qualitative and quantitative characteristics are presented. A number of regression models combining elements of classical and fuzzy regressions are presented.

All methods and models developed by the authors and described in the book are illustrated with examples from various fields of human activities.

This book meant for scientists in the field of computer science, expert systems, artificial intelligence and decision making; and also for engineers, post-graduate students and students who study the fuzzy set theory and its applications.

68 Tables. 25 Illustrations. References - 230 items.

Olga Poleshchuk
Evgeniy Komarov

Introduction

While solving the problems in various areas of activity (engineering, ecology, education, medicine, economy etc.) a person can be both an observer of processes going on in these areas, or a directly involved participant (expert), i.e. he/she can influence the processes through his/her subjective opinion, knowledge and experience. With that, being an expert, a person brings a subjective component to the processes described or estimated that should be considered. The information coming from various experts can contain both accurate and fuzzy data. Objectively, the latter prevail because a person uses words of a natural language to estimate processes, events and objects. For example, manufacturability — low, interface — friendly, qualification — high, compliance — complete, probability — high, etc.

Expert information with fuzzy data (fuzzy expert information) is difficult to formalize within the limits of traditional mathematical concepts. Numerical characteristics feature intermittent transitions from one linguistic value to another, as a result objects with boundary physical values of these characteristics are difficult to describe. While mapping linguistic values of qualitative characteristics to numerical elements of ordinal scales the information is coarsened, and its valuable component characterizing individual experience and knowledge of a person, gets lost. Attempts to formalize fuzzy information on the basis of classical and judgmental probabilities were not success because of known restrictive requirements applied to their use.

The model approach based on the fuzzy set theory has allowed to eliminate these shortages of traditional formalizations of fuzzy information. Successful development of the fuzzy set theory has ensured its recognition; however it has revealed points, which need to be solved. A section of the fuzzy set theory, which permanently faces criticism, is the stage dealing with obtained data formalization or building of relevant membership functions. As a rule, requirements to formalization models are formulated within the scope of each specific problem, and quality of built-up models often depends on experience and skill of contributors. Apparently, a reason of such dependence is that formalization methods are limited by both a type of the information and the ways experts provide it.

In our book, semantic spaces with completeness and orthogonality properties (complete orthogonal semantic spaces) are considered as models of expert evaluations of objects. Theoretical and practical studies of some Russian and non-Russian scientists have shown that these models describe experts' evaluations best of all, and as a result they are included rather often in more sophisticated models used in intellectual systems for decision making and data analysis.

Chapter 1 contains information related to the fuzzy set theory, cluster and regressive analyses. Chapter 2 is devoted to techniques of constructing complete orthogonal semantic spaces.

It is obvious, that models of formalization of the expert information obtained from different sources can differ. As a rule, while working out fuzzy models, membership functions of formalized concepts are updated and adjusted in the course of the real analysis of system behavior. To reduce time and material expenditures needed to determine these functions, the authors have developed methods of comparative analysis of such models, studies of structural composition of their sets and techniques for constructing generalized models. Chapters 3 and 4 are devoted to these methods.

Chapter 5 is devoted to obtaining rating points of qualitative and quantitative characteristics of objects and groups of objects. The basic problem of obtaining rating points is that while using ordinal scales to estimate characteristics, false arithmetical operations are employed sometimes, which lead to unstable final results. Methods of expert information formalization developed in Chapter 2 allow unifying characteristics and correctly handling dimensionless generalized quantities, which are values of their membership functions, rather than the values of characteristics.

Chapter 6 is devoted to fuzzy regression analysis methods which have considerably expanded a range of application of known classical regression analysis methods. However, limited range of input data membership functions in the fuzzy regression analysis methods caused a gap which is partially bridged here. Two models which combine elements of classical and fuzzy regression models are presented. Input and output data of these models are values of complete orthogonal semantic spaces.

All methods and the models developed by the authors are illustrated by examples taken from various fields of human activities.

Historically a practical component of the fuzzy set theory is developed much faster than the theoretical one, thus the book draws attention not only to practical applications based on new techniques developed by the authors, but also to the theoretical aspects. According to the authors, such an approach of presentation allows contributors to further develop these methods, and to successfully put them into practice.

Olga Poleshchuk
Evgeniy Komarov

Contents

Chapter 1
The Basic Concepts of the Fuzzy Set Theory

1.1 Uncertainty Classification

The common feature of problems solvable with active participation of experts is presence of the versatile information which is difficult for mathematical formalization within the scope of traditional concepts. It is partly connected with the fact that while describing or estimating real objects and situations experts use words of a natural language to reflect their subjective experience, judgment, sights and interests. Use of such words brings uncertainty to information obtained from an expert in the form of fuzziness. In usual languages the concepts of uncertainty, fuzziness and randomness tend to mix up, however they were differentiated in a language of science many years ago. With this assumption, to review such philosophical concept as uncertainty, the following classification [1] illustrated by Fig. 1.1 is recommended to use.

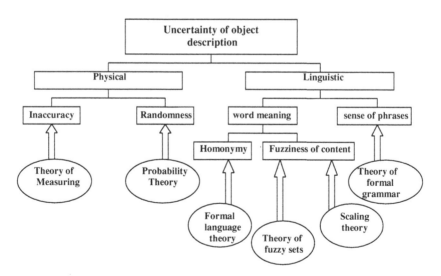

Fig. 1.1 Classification of object description uncertainties

O.M. Poleshchuk and E.G. Komarov: Expert Fuzzy Info. Processing, STUDFUZZ 268, pp. 1–41.
springerlink.com © Springer-Verlag Berlin Heidelberg 2011

Physical uncertainty originates from description of the real world from a point of view of an observer. The theory of measuring [2] deals with an imprecision (or inaccuracy), and the probability theory deals with randomness [3]. Linguistic uncertainty includes uncertainties of natural language concepts and structures. The theory of formal grammars deals with ambiguity of sense of phrases. Uncertainty of word meanings is handled by theories of fuzzy sets [4] and scaling [2].

Before the study of organizational problems connected with human decision-making and building of automated control systems accompanied with simulating of human operator activity started, handling and analysis of information featuring physical uncertainty were successfully carried out by methods of the classical probability theory and mathematical statistics, and elements of control systems were represented within the scope of the classical theory of sets. The probability theory was the first theory which deals with uncertainty, more precisely, with its variety, which is randomness; consequently, among all similar theories it has been better detailed. However, the probability theory is based on a number of requirements [3], the satisfaction of which is necessary to ensure adequacy of conclusions resulted from analysis of the information formalized by its methods [5—7]. Examples of such requirements are given below:

- guarantees that the observable attributes can be spread to all objects or events of the certain type (population);
- unlimited recurrence (repeatability) of observable attributes under identical conditions;
- independence of events, etc.

From the very beginning of their history, the methods of probability theory and mathematical statistics addressed the study of physical uncertainties (randomness). If information is obtained under conditions either artificially created or observer-independent [8—9], one can speak about adequacy of application of probability theory and mathematical statistics methods for the purposes of handling and analysis of such information [10—12]. In real situations these conditions are seldom met. Standard probability theory is a more than adequate tool when one deals with physical, that is, inanimate systems in which human judgment, perceptions and emotions do not have a role. But there is another realm, the realm of what are called humanistic systems--systems in which human judgment, perceptions and emotions play a prominent role. This is the realm of economics, education, psychology, law and decision analysis.

For expert evaluations, objectively an opportunity to provide identical conditions for repeated observable attributes is rather small as the opinions, judgments of experts and the level of their competence introduce a subjective factor into evaluation procedure, and this factor changes conditions of the estimating procedure implementation, and thus shall be considered.

So, classical probabilities were tried to be substituted with non-classical ones [13]. Among non-classical probabilities, valence and axiological probabilities are most often used. The valence probability expresses expectancy of hypothesis realization taking into account a context of actual demonstrations on an object of research. In a particular case, when the representative sampling is a sampling of homogeneous events, the valence probability is a statistical (classical) one. The axiological probability expresses expectancy of hypothesis realization taking into account a context of value judgments advanced by one or several experts regarding an object of research. When using such probability, risk of arbitrariness and erroneous prognosis of object behavior increases invariably. Besides, analyses of expert evaluations showed that thoughts and judgments of experts are non-additive, which means that measures used by experts are free from additivity property which is inherent to a probability measure.

Attempts to formalize expert logical-linguistic expressions and value judgments of observable properties of a certain object resulted in led emergence of the fuzzy set theory developed by L. Zadeh. In 1965 in the "Information and Control" journal he published his "Fuzzy Sets" paper [14] which became a powerful impetus for theoretical evidences and applied researches in various areas with active participation of human experts.

L. Zadeh relied on a premise saying that outcome of human thinking is not a set of numbers, but elements of some fuzzy sets or classes of objects, for which transition from "membership" to "non-membership" is not intermittent, but continuous. He understood that use of such objects is a means to improve stability of mathematical models of real human activity events.

Distinction between fuzziness and randomness leads to the fact that methods of the fuzzy set theory are not similar to probability theory methods, they are simpler in many aspects, because they are based on the simpler concept of membership function (by L. Zadeh) in comparison with a probability distribution function which assumes definition of a probability measure generating this function [15].

Obviously, the information obtained from experts can contain both accurate and fuzzy data. Fuzzy data arise from use of linguistic values of estimated characteristics within the scope of professional language of experts. The information containing fuzzy data was called the fuzzy expert information.

The history of the fuzzy set theory can give some examples of the unsuccessful practical applications caused by badly considered adherence to the new scientific direction. That is why it is necessary to define a class of problems for which its use is expedient. This class is characterized with complexity of quantitative evaluation of objects and processes considered, presence of information difficult to formalize accompanied with uncertainty of nonrandom nature, necessity of accounting for individual characteristics and peculiarities of persons (experts) in charge of making evaluations.

1.2 Scales and Admissible Transformations

Let us consider known scales and admissible transformations of characteristic values measured within these scales. It is known that a transformation which keeps substantial sense of the given aspect of measurement [16] is called an admissible transformation of the measured characteristic values. To measure quantitative characteristics the following scales are used, in particular, absolute, ratio, interval, difference.

The values of quantitative characteristics measured with these scales are referred to as physical ones.

To measure qualitative characteristics the following scales are used, in particular, names, ordinal (rank).

The research conclusions made can be adequate to a reality if and when they do not depend on what unit of measurement is preferred by a researcher, i.e. these conclusions should be invariant with regard to admissible transformation of the characteristic values measured in any scale. Let us give admissible transformations of function $\Phi(x)$ in the scales below [17—20]:

Absolute scale	$\Phi(x) = x$
Ratio scale	$\Phi(x) = ax \ \ (a > 0)$
Scale of intervals	$\Phi(x) = ax + b \ \ (a > 0; \ b \in R)$
Scale of differences	$\Phi(x) = x + b \ \ (b \in R)$
Scale of names	$\Phi(x)$ — all one-to-one transformations
Ordinal (rank) scale	$\Phi(x)$ — all strictly increasing transformations

When experts use ordinal scales to measure qualitative characteristics, then for definition of aggregating indicators average values of score expert evaluations are used often enough [21—27]. There are some methods of average computing, in particular, arithmetic mean, geometrical mean, harmonic mean, mean square value, mode, median. Let us consider application of an arithmetic mean in an ordinal scale, being most often used. Let us assume that two entrants got marks 4 and 3, accordingly, for the one entrance examination, and marks 4 and 5 for the other entrance examination. Their total scores and arithmetic mean values of two examinations are identical and equal to 8 and 4, accordingly. The conclusion is that they have equal rights to matriculation. As examination scores are allotted according to an ordinal scale, let us use strictly increasing transformation of this scale Φ: $\Phi(3) = 3$; $\Phi(4) = 4$; $\Phi(5) = 7$. According to the transformation made (which is admissible), the total score and the arithmetic mean of marks of one entrant remained unchanged, and the same of the second entrant became equal to 10 and 5, accordingly. Thus, the second entrant has preferable rights for matriculation than the first one. With the admissible transformation completed, the result stability is broken that means incorrectness of arithmetical operations in the ordinal and nominal scales [28].

Having in mind that used of averages in various scales is spread enough, the problems [29—32] of achieving average values were set and solved, with results of the average comparison being stable with regard to the admissible transformations of characteristic values measured in the specific scale. Let u provide definitions of mean values according to Kolmogorov and Cauchy.

For numbers $x_1, x_2, \ldots x_n$, a mean value according to Kolmogorov is the following function

$$F^{-1}\left(\frac{F(x_1) + F(x_2) + \ldots + F(x_n)}{n}\right),$$

where $F(x)$ is strictly monotone function; $F^{-1}(x)$ is inverse function to $F(x)$.

If $F(x) = x$, the average according to Kolmogorov is an arithmetic mean, if $F(x) = \ln x$, it is an geometrical mean, if $F(x) = 1/x$ it is a harmonic mean.

It is proven [18] that according to Kolmogorov, among all mean values only an arithmetic mean is possible to use in a scale of intervals, and only power means and a geometrical mean are possible to use in a scale of ratios.

For numbers $x_1, x_2, \ldots x_n$, according to Cauchy, a mean value is the function $f(x_1, x_2, \ldots x_n)$, if $\min(x_1, x_2, \ldots x_n) \leq f(x_1, x_2, \ldots x_n) \leq \max(x_1, x_2, \ldots x_n)$.

It is proven [18] that according to Cauchy, among all mean values only terms of a variation series, in particular, a median, can be used in an ordinal scale. Use of terms of a variation series to determine an aggregating indicator is often noninformational for a number of particular indicators owing to very coarse estimate. For example, they are used in educational process, where knowledge is estimated in marks from two to five.

To estimate qualitative and quantitative characteristics, experts use verbal scales often enough. Values in verbal scales are words expressing characteristic appearance intensity degree. These words are referred to as levels or gradations. Let us consider only those verbal scales with which it is possible to define a linear order, i.e. "less — more" ratio.

Problems of definition of sets of verbal scale levels and quantitative values of qualitative characteristics appearance within the limits of these levels are main ones in expert evaluations [33]. For the purpose of employment of known mathematical models of information processing, numerical points are put in correspondence to levels of verbal scales. The result of this approach is that the verbal scale is mapped to a verbally-numerical scale. Definition of values of the points put in correspondence to levels of verbal scales is a separate problem, the solution of which influences the stability of the results obtained within the limits of a mathematical model, so the justification of use of these values is needed. For example, marks "2", "3", "4", "5", which are put in correspondence to verbal values E ("unsatisfactory"), C ("satisfactory"), B ("good"), A ("excellent"), compose a verbally-numerical scale in their aggregate. Certainly, it is ought to remember that the numbers put in correspondence to verbal levels of qualitative characteristics are elements of an ordinal scale and all restrictions mentioned above are applicable to them.

However, if within the scope of a specific problem use of a certain verbally-numerical scale is justified, in actual practice experts face essential difficulties caused by intermittent transitions between levels, not allowing to catch and estimate intermediate conditions of the characteristic under evaluation [34]. To estimate intermediate conditions, process of artificial fuzzification of numerical points corresponding to levels of verbal scales is applied. For example, in educational process when evaluating the pupils' knowledge without any limitations imposed on generality of "good" knowledge, not only mark "4", but also the whole range of marks [3.5; 4.5] is quite often used. Such process of points fuzzification simulates smoothness of estimating activity of experts, but does not facilitate process of exposing real objects with evaluations arranged near the boundaries of fuzzy areas.

Verbal scales are used often enough to describe physical values of quantitative characteristics. For example, in [35. 36], to describe a "steam pressure at inlet" (with a range [1.1; 6.7]) parameter of a high pressure preheater intended to improve turbine plant efficiency, the verbal scale with levels "low pressure", "pressure close to 4", 'high pressure" is used. Another example is the verbal scale used to describe event probabilities. As known, an event probability is usually expressed by numerical value and varies from 0 to 1. However, for example, when speaking about the probability of an enterprise bankruptcy, the manager of this enterprise is interested not in a precise figure, which is likely to be little-informative for him, but in definition of one of verbal levels of bankruptcy probability, in particular, "very low", "low", "mean", "high", "very high" [37].

With a range of definition (universal set) of quantitative characteristic and levels of a verbal scale known, an expert divides this area into non-overlapping sets which correspond to verbal levels. However, such approach is featuring with essential shortage which lies in the fact that while describing objects with boundary values of an indicator, an expert experiences difficulties caused by intermittent transitions between values.

This shortage can be remedied with the fuzzy set theory in which not precise intervals of values are put in correspondence to verbal levels of quantitative property, but fuzzy sets. The resultant verbal-fuzzy scale is referred to as a linguistic scale [38—39]. As a result of such buildings, a quantitative characteristic, on the one hand, is corresponded with physical values measured by a technical instrument and, on the other hand, with linguistic values "measured" by an expert. Each physical value belongs to some linguistic one with certain degree of expert confidence.

Building of a linguistic scale for qualitative characteristics is much more complicated. If a verbal-numerical scale for qualitative characteristics represents a collection of verbal levels with the corresponded collection of numbers (elements of an ordinal scale), then a linguistic scale is a collection of verbal levels with a collection of the corresponded fuzzy sets specified at some universe. As qualitative characteristics cannot be measured objectively (by instrument), the universal sets applicable for them cannot be unambiguously defined, as they do for quantitative characteristics. Definition of universal set is made within the scope of each qualitative characteristic and requirements of each specific task.

Thus, expedient values of linguistic scales for qualitative characteristics are fuzzy sets. In the mathematical statistics, collections of numerical data and corresponded chance quantities are referred to as sampling; similarly, in the fuzzy set theory verbal levels and corresponded fuzzy sets are referred to as linguistic values.

Definition of linguistic values of characteristics (based on the fuzzy set theory) makes it possible to operate not with values of the characteristics which are non-comparable among themselves by substance and content (as they are estimated in different scales and having different dimensions), but with dimensionless values of membership functions.

1.3 Fuzzy Sets

Though the concept of a set plays a fundamental role not only in the mathematics science, it has no rigorous definition. It is considered that this word is commonly understood as a quantity of roughly homogeneous (in any sense) elements.

Within this set it is possible to define subsets and it is necessary to do this strictly and explicitly. Let us have a set X consisting of elements x. Its subset A can be defined, for example, by means of characteristic function

$$h_A(x) = \begin{cases} 1, & x \in A; \\ 0, & x \notin A. \end{cases}$$

Thus, characteristic function allows mapping of a set to another set of two elements: 0 and 1. It can be written as follows:

$$h_A(x): X \rightarrow \{0,1\}.$$

So, for any element of set X there are two possibilities: it can either belong or not belong to a set A.

If to consider a set of all subsets, it is possible to apply, in a certain way, operations of intersection, union and complement to this set, and those can be expressed as operations to corresponding characteristic functions.

Using sets, we can define various concepts. Let us explain with the following example.

Example 1.1. Formalization of the "normal functioning of an object" concept on the basis of the classical theory of sets. Let some parameter be defined at a universal set and takes values from X_m to X_M. Then, based on the substantial sense of a problem, the concept of "normal functioning of an object" can be defined by means of a set A the characteristic function of which is shown in Fig. 1.2.a.

Under this formalization it is assumed that all users unambiguously understand the given concept and agree with boundary values $\overline{x_m}$ and $\overline{x_M}$. This assumption is met, for example, when boundary values are computed on the basis of an approved mathematical model of the process of an investigated object functioning.

Operations of intersection, union and complement can be interpreted as logical connectives "and", "or" and "not", accordingly. In this case, we mean Boolean (two-value, binary) logic. With expressions of some concepts $a_1, a_2, ..., a_n$ in the form of sets $A_1, A_2, ..., A_n$ available, it is possible to find the sets corresponding to these concepts, in the form of logic functions $f(a_1, a_2, ..., a_n)$.

The theory of sets and corresponding Boolean logic makes the foundation of classical mathematics and everything based on it, up to advanced computer processors. Models of complicated engineering and physical systems, and chemical processes were well described with this language and successfully implemented using computers. Insufficient speed, memory size, and also some other engineering complexities of implementation of these models were the only problem.

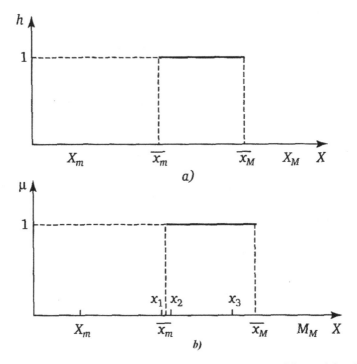

Fig. 1.2 Characteristic function of A set, which formalizes concept of "normal functioning of an object": a — by means of the theory of sets; b — by means of human analysis of a situation

The situation changed dramatically with the necessity to consider peculiarities of human perception, evaluation and analysis of the information as equal part of a simulated system. Let us consider an example 1.1 (Fig. 1.2. b) from the point of view of the description of a situation made by a human.

Example 1.1 (cont'd). Let us consider three values: x_1, x_2 and x_3 shown in Fig. 1.2.b.

Certain paradox is obvious: in the model values x_1 and x_2 are different and x_2 and x_3 are identical (with regard to the "normal functioning of an object" concept formalized by means of set A). Thus, the computer model can "see" actually close situations as different ones, and situations different physically as identical ones. If there is no unambiguous rule (model) to compute boundary values $\overline{x_m}$ and $\overline{x_M}$, then considered situation can lead to the situation when the conclusions derived from the analysis of such models, will not correspond to perceptions of experts.

This discordance between the theory of sets language and a human mode of thinking is one of causes of unsuccessful attempts in use of mathematical methods and computer technologies in those fields of activity where influence of the human factor is significant.

The solution of the problem was offered by L. Zadeh in 1965. In [14] he introduced the concept of a fuzzy set.

The basic idea of Zadeh consisted in "allowing" a characteristic function to accept not only value 0 (complete non-membership) or 1 (complete membership), but also intermediate values of a membership from a segment [0. 1]. Thus, he has substituted the concept of characteristic function with the concept of membership function

$$\mu_{\tilde{A}}(x): X \to [0,1]$$

According to [14], the set of pairs of the following form is referred to as a fuzzy set $\tilde{A}$

$$\{[x, \mu_{\tilde{A}}(x)] : x \in X\}$$

From the definition one can understand that specification of a fuzzy subset $\tilde{A}$ in X is equivalent to specification of its membership function $\mu_{\tilde{A}}(x)$. Following the traditional way, we will use the term "fuzzy set" instead of more correct term "fuzzy subset".

Value of membership function $\mu_{\tilde{A}}(x)$ for an element x to fuzzy set $\tilde{A}$ is referred to as grade of membership x to $\tilde{A}$. This value can be interpreted as the level of an element correspondence to the concept formalized by a fuzzy set $\tilde{A}$, with the correspondence level being determined by an expert (group of experts).

Domain of a membership function $\mu_{\tilde{A}}(x)$ is referred to as universal set X of fuzzy set $\tilde{A}$.

A set which membership function is equal to zero for all elements of universal set X is referred to as empty set $\varnothing$).

Using fuzzy sets, it is possible to define various concepts in more natural manner of perception and description of objects. Let us explain with an example.

Example 1.2. Formalization of the concept of "normal functioning of an object" on the basis of the fuzzy set theory. As in the example 1.1., let a parameter x be defined at universal set X and take values from X_m to X_M. The concept of "normal functioning of an object" can be defined as fuzzy set $\tilde{A}$ which membership function is shown in Fig. 1.3.

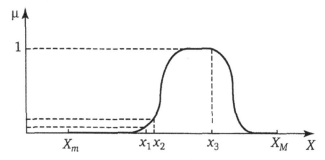

Fig. 1.3 Membership function of a fuzzy set $\tilde{A}$ formalizing the concept of "normal functioning of an object"

By comparing Fig. 1.2.b and Fig. 1.3., one can notice the collapse of the paradox highlighted when analyzing the example 1.1: being formalized by means of fuzzy sets, the model recognizes values x_1 and x_2 are seen as close ones. Thus, the computer model based on principles of the fuzzy logic, "perceives" actually close situations as similar ones.

It is worth mentioning that there is a tendency of probability treatment of fuzzy set [20]. Obviously there is no sense to compare concepts of probability and fuzzy set at the same abstracting level. Let us give more details about this.

Let (X,B,P) is a probability space [3]: $B \subseteq P(X)$ is a field of Borel subsets of set X, $B \rightarrow [0,1]$ is the probability measure P, which satisfies to the conditions:

1) $A \subseteq X \Leftrightarrow P(A) \geq 0$;
2) $P(\phi) = 0$;
3) if $A, B \in B$, then $P(A \cup B) = P(A) + P(B) - P(A \cap B)$.

According to L. Zadeh, fuzzy set is described by membership function $\mu(x)$ taking its values in a point of the segment [0.1]. From the point of view of the mapping theory, $P : B \rightarrow [0,1]$ and $\mu : X \rightarrow [0,1]$ are absolutely different objects. Probability p is defined with σ-algebra B, and $\mu(x)$ is a usual function with a range of definition X.

So, it is possible to draw some analogies between the membership function of fuzzy set and probability density function of a chance quantity, but not to identify them.

A clear subset of universal set, defined as

$$A_\alpha = \{x \in X : \mu_{\tilde{A}}(x) \geq \alpha\}, \ \alpha \in [0,1], \tag{1.1}$$

is referred to as set of α-level (α-cut) of fuzzy set $\tilde{A}$ with membership function $\mu_{\tilde{A}}(x)$).

Example 1.3. Definition of α-level sets. Let $\tilde{A} = \{0,3/x_1; 0,7/x_2; 0,9/x_3; 0,5/x_4\}$, then $A_{0,4} = \{x_2, x_3, x_4\}$; $A_{0,7} = \{x_2, x_3\}$; $A_{0,9} = \{x_3\}$.

It is obvious, that with $\alpha_1 \geq \alpha_2$ condition $A_{\alpha_1} \geq A_{\alpha_2}$ is satisfied.

The Theorem of Decomposition [40]. Any fuzzy set $\tilde{A}$ with membership function $\mu_{\tilde{A}}(x)$ can be decomposed by α-level sets

$$\tilde{A} = \bigcup_{\alpha \in [0,1]} \alpha A_\alpha \tag{1.2}$$

or

$$\mu_{\tilde{A}}(x) = \bigcup_{\alpha \in [0,1]} [\alpha \mu_{A_\alpha}(x)],$$

where

$$\mu_{A_\alpha}(x) = \begin{cases} 1, & \mu_{\tilde{A}}(x) \geq \alpha; \\ 0, & \mu_{\tilde{A}}(x) < \alpha. \end{cases}$$

The decomposition forms the basis for a method of fuzziness formalization. According to this method, the fuzziness is expressed by means of a collection of hierarchically ordered definite sets.

Let us define the basic set-theoretic operations at set $H(X)$ of all fuzzy subsets of a definite set X.

Fuzzy sets $\tilde{A}$ and $\tilde{B}$ are equal, if for all $x \in X$ the condition $\mu_{\tilde{A}}(x) = \mu_{\tilde{B}}(x)$ is satisfied.

The fuzzy set $\tilde{A}$ belongs to fuzzy set $\tilde{B}$, $\tilde{A} \subseteq \tilde{B}$, if $\mu_{\tilde{A}}(x) \leq \mu_{\tilde{B}}(x)$; $\forall x \in X$.

The fuzzy set $\overline{\tilde{A}}$ is referred to as complement of fuzzy set $\tilde{A}$, if

$$\mu_{\overline{\tilde{A}}}(x) = 1 - \mu_{\tilde{A}}(x); \ \forall x \in X.$$

The fuzzy set $\tilde{C}$ is referred to as intersection of fuzzy sets $\tilde{A}$ and $\tilde{B}$, $\tilde{C} = \tilde{A} \cap \tilde{B}$, if $\mu_{\tilde{C}}(x) = \mu_{\tilde{A}}(x) \wedge \mu_{\tilde{B}}(x)$; $\forall x \in X$, where $\wedge$ is an operator from triangular norm class.

Valid binary function $T : [0,1] \times [0,1] \to [0,1]$ is referred to as triangular norm, this function satisfies the following conditions:

1. $T(0,0) = 0, T(\mu_{\tilde{A}}, 1) = T(1, \mu_{\tilde{A}}) = \mu_{\tilde{A}}$ (boundedness);
2. $T(\mu_{\tilde{A}}, \mu_{\tilde{B}}) \leq T(\mu_{\tilde{C}}, \mu_{\tilde{D}})$, if $\mu_{\tilde{A}} \leq \mu_{\tilde{C}}$; $\mu_{\tilde{B}} \leq \mu_{\tilde{D}}$ (monotonicity property);
3. $T(\mu_{\tilde{A}}, \mu_{\tilde{B}}) = T(\mu_{\tilde{B}}, \mu_{\tilde{A}})$ (commutativity);
4. $T[\mu_{\tilde{A}}, T(\mu_{\tilde{B}}, \mu_{\tilde{C}})] = T[T(\mu_{\tilde{A}}, \mu_{\tilde{B}}), \mu_{\tilde{C}}]$ (associativity).

The pair $([0,1], T)$ forms a commutative semigroup with a unity.
Examples of triangular norms are:

1. $T(\mu_{\tilde{A}}, \mu_{\tilde{B}}) = \min(\mu_{\tilde{A}}, \mu_{\tilde{B}})$;
2. $T(\mu_{\tilde{A}}, \mu_{\tilde{B}}) = \mu_{\tilde{A}} \times \mu_{\tilde{B}}$;
3. $T(\mu_{\tilde{A}}, \mu_{\tilde{B}}) = \max(0, \mu_{\tilde{A}} + \mu_{\tilde{B}} - 1)$.

The fuzzy set $\tilde{C}$ is referred to as union of fuzzy sets $\tilde{A}$ and $\tilde{B}$, $\tilde{C} = \tilde{A} \cup \tilde{B}$, if $\mu_{\tilde{C}}(x) = \mu_{\tilde{A}}(x) \vee \mu_{\tilde{B}}(x)$; $\forall x \in X$.

A real-valued binary function is referred to as triangular conorm, $K : [0,1] \times [0,1] \to [0,1]$, if this function satisfies to following conditions:

1. $K(0,0) = 0, K(\mu_{\tilde{A}}, 1) = K(1, \mu_{\tilde{A}}) = \mu_{\tilde{A}}$ (boundedness);
2. $K(\mu_{\tilde{A}}, \mu_{\tilde{B}}) \leq K(\mu_{\tilde{C}}, \mu_{\tilde{D}})$, if $\mu_{\tilde{A}} \leq \mu_{\tilde{C}}, \mu_{\tilde{B}} \leq \mu_{\tilde{D}}$ (monotonicity property);
3. $K(\mu_{\tilde{A}}, \mu_{\tilde{B}}) = K(\mu_{\tilde{B}}, \mu_{\tilde{A}})$ (commutativity);
4. $K[\mu_{\tilde{A}}, K(\mu_{\tilde{B}}, \mu_{\tilde{C}})] = K[K(\mu_{\tilde{A}}, \mu_{\tilde{B}}), \mu_{\tilde{C}}]$ (associativity).

Examples of triangular conorms are:

1. $K(\mu_{\tilde{A}}, \mu_{\tilde{B}}) = \max(\mu_{\tilde{A}}, \mu_{\tilde{B}})$;
2. $K(\mu_{\tilde{A}}, \mu_{\tilde{B}}) = \mu_{\tilde{A}} + \mu_{\tilde{B}} - \mu_{\tilde{A}} \times \mu_{\tilde{B}}$;
3. $K(\mu_{\tilde{A}}, \mu_{\tilde{B}}) = \min(1, \mu_{\tilde{A}}, \mu_{\tilde{B}})$

Let us obtain following definitions for triangular norm (min), and for triangular conorm (max).

The fuzzy set $\tilde{C}$ is referred to as intersection of fuzzy sets $\tilde{A}$ and $\tilde{B}$, $\tilde{C} = \tilde{A} \cap \tilde{B}$, if

$$\mu_{\tilde{C}}(x) = \min[\mu_{\tilde{A}}(x), \mu_{\tilde{B}}(x)], \ \forall x \in X.$$

The fuzzy set $\tilde{C}$ is referred to as union of fuzzy sets $\tilde{A}$ and $\tilde{B}$, $\tilde{C} = \tilde{A} \cup \tilde{B}$, if

$$\mu_{\tilde{C}}(x) = \max[\mu_{\tilde{A}}(x), \mu_{\tilde{B}}(x)], \ \forall x \in X.$$

The fuzzy set $\tilde{C}$ is referred to as a difference of fuzzy sets $\tilde{A}$ and $\tilde{B}$, $\tilde{C} = \tilde{A} - \tilde{B} = \tilde{A} \cup \tilde{B}$, if

$$\mu_{\tilde{C}}(x) = \min[\mu_{\tilde{A}}(x), 1 - \mu_{\tilde{B}}(x)], \ \forall x \in X.$$

The fuzzy set $\tilde{C}$ is referred to as a disjunctive sum of fuzzy sets $\tilde{A}$ and $\tilde{B}$, $\tilde{C} = \tilde{A} \oplus \tilde{B} = (A - B) \cup (B - A)$, if

$$\mu_{\tilde{C}}(x) = \max[\min(\mu_{\tilde{A}}(x), 1 - \mu_{\tilde{B}}(x)), \min(1 - \mu_{\tilde{A}}(x), \mu_{\tilde{B}}(x))],$$
$$\forall x \in X$$

Example 1.4. Operations with fuzzy sets. Let
$$\tilde{A} = \{0,3 / x_1; 0,7 / x_2; 0,9 / x_3; 0,5 / x_4\};$$
$$\tilde{B} = \{0,1 / x_1; 0,8 / x_2; 1 / x_3; 0,2 / x_4\}.$$

Then
$$\overline{\tilde{A}} = \{0,7 / x_1; 0,3 / x_2; 0,1 / x_3; 0,5 / x_4\};$$
$$\overline{\tilde{B}} = \{0,9 / x_1; 0,2 / x_2; 0 / x_3; 0,8 / x_4\};$$

$$\tilde{A} \cap \tilde{B} = \{0,1 / x_1; 0,7 / x_2; 0,9 / x_3; 0,2 / x_4\};$$
$$\tilde{A} \cup \tilde{B} = \{0,3 / x_1; 0,8 / x_2; 1 / x_3; 0,5 / x_4\};$$
$$\tilde{A} - \tilde{B} = \tilde{A} \cap \overline{\tilde{B}} = \{0,3 / x_1; 0,2 / x_2; 0 / x_3; 0,5 / x_4\};$$
$$\tilde{B} - \tilde{A} = \tilde{B} \cap \overline{\tilde{A}} = \{0,1 / x_1; 0,3 / x_2; 0,1 / x_3; 0,2 / x_4\};$$
$$\tilde{A} \oplus \tilde{B} = \{0,3 / x_1; 0,3 / x_2; 0,1 / x_3; 0,5 / x_4\}$$

Let us consider properties of operations $\cup$ and $\cap$. Let $\tilde{A}$, $\tilde{B}$, $\tilde{C}$ be fuzzy sets, then following conditions are satisfied.

1. Commutativity

$$\tilde{A} \cup \tilde{B} = \tilde{B} \cup \tilde{A}, \ \tilde{A} \cap \tilde{B} = \tilde{B} \cap \tilde{A}.$$

2. Associativity

$$\left(\tilde{A} \cap \tilde{B}\right) \cap \tilde{C} = \tilde{A} \cap \left(\tilde{B} \cap \tilde{C}\right), \ \left(\tilde{A} \cup \tilde{B}\right) \cup \tilde{C} = \tilde{A} \cup \left(\tilde{B} \cap \tilde{C}\right).$$

3. Idempotency

$$\tilde{A} \cap \tilde{A} = \tilde{A}, \ \tilde{A} \cup \tilde{A} = \tilde{A}.$$

4. Distributivity

$$\tilde{A} \cap \left(\tilde{B} \cup \tilde{C}\right) = \left(\tilde{A} \cap \tilde{B}\right) \cup \left(\tilde{A} \cap \tilde{C}\right), \ \tilde{A} \cup \left(\tilde{B} \cap \tilde{C}\right) = \left(\tilde{A} \cup \tilde{B}\right) \cap \left(\tilde{A} \cup \tilde{C}\right).$$

5. $\tilde{A} \cup \phi = \tilde{A}.$

6. $\tilde{A} \cap \phi = \phi.$

7. $\tilde{A} \cup X = X.$

8. $\tilde{A} \cap X = \tilde{A}.$

9. Laws of dualization

$$\overline{\tilde{A} \cap \tilde{B}} = \overline{\tilde{A}} \cup \overline{\tilde{B}}, \ \overline{\tilde{A} \cup \tilde{B}} = \overline{\tilde{A}} \cap \overline{\tilde{B}}.$$

Unlike definite sets, for fuzzy ones, in the general case:

$$\tilde{A} \cap \overline{\tilde{A}} \neq \phi, \ \tilde{A} \cap \overline{\tilde{A}} \neq \phi.$$

There is a possibility to construct methods of processing and analysis of fuzzy information with use of fuzzy sets and arithmetical operations [15. 41—44]. The algebra of fuzzy numbers is the mathematical basis to construct such methods.

Fuzzy number (FN) $\tilde{A}$ is referred to as fuzzy subset of set of real numbers R possessing membership function

$$\mu_{\tilde{A}} : R \rightarrow [0,1].$$

The subset $S_A \subset R$ is referred to as the support of FN $\tilde{A}$, if

$$S_A = \text{supp} A = \left[x : \mu_{\tilde{A}}(x) > 0\right].$$

FN $\tilde{A}$ is referred to as positive, if $\forall x \in S_A$, $x > 0$, and negative, if $\forall x \in S_A$, $x < 0$.

A subset of the real line R, which defined in the form

$$A_\alpha = \{x \in R : \mu_{\tilde{A}}(x) \geq \alpha\}, \quad \alpha \in [0,1] \qquad (1.3)$$

is referred to as set of α-level (α-cut) of FN $\tilde{A}$ with membership function $\mu_{\tilde{A}}(x)$, by analogy with (1.1).

FN $\tilde{A}$ with membership function $\mu_{\tilde{A}}(x)$ is referred to as normal, if

$$\max_{x} \mu_{\tilde{A}}(x) = 1, \quad x \in R.$$

FN $\tilde{A}$ with membership function $\mu_{\tilde{A}}(x)$ is referred to as unimodal, if there is a unique point $x \in R$, for which the equality $\mu_{\tilde{A}}(x) = 0$ is satisfied.

FN $\tilde{A}$ with membership function $\mu_{\tilde{A}}(x)$ is referred to as multimodal, if a point $x \in R : \mu_{\tilde{A}}(x) = 1$ is not unique.

FN $\tilde{A}$ with membership function $\mu_{\tilde{A}}(x)$ is referred to as tolerant, if there is an interval for all points of which the equality $\mu_{\tilde{A}}(x) = 1$ is satisfied. This interval is referred to as an interval of tolerance of FN $\tilde{A}$.

The expanded binary arithmetical operation denoted as $\tilde{\nabla}$ [15], for fuzzy numbers $\tilde{A}$, $\tilde{B}$ with membership functions $\mu_{\tilde{A}}(x)$, $\mu_{\tilde{B}}(x)$, accordingly, is defined as follows:

$$\tilde{C} = \tilde{A}\tilde{\nabla}\tilde{B} \Leftrightarrow \mu_{\tilde{C}}(z) = \bigvee_{z=x\nabla y} \left[\mu_{\tilde{A}}(z) \wedge \mu_{\tilde{B}}(y)\right], \forall x, y, z \in R. \qquad (1.4)$$

Based on (1.4), one can define such arithmetical operations as expanded addition, subtraction, multiplication and division of FN $\tilde{A}$, $\tilde{B}$ with membership functions $\mu_{\tilde{A}}(x)$, $\mu_{\tilde{B}}(x)$, accordingly, for special cases of an intersection operator $\wedge$ in a class of triangular norms $\mu_{\tilde{A}}(x) \wedge \mu_{\tilde{B}}(x) = \min(\mu_{\tilde{A}}, \mu_{\tilde{B}})$, and also an union operator $\vee$ in a class of triangular conorms $\mu_{\tilde{A}}(x) \vee \mu_{\tilde{B}}(x) = \min(\mu_{\tilde{A}}, \mu_{\tilde{B}})$:

$$\left.\begin{array}{l} \tilde{C} = \tilde{A} \oplus \tilde{B} \Leftrightarrow \mu_{\tilde{C}}(z) = \max_{z=x+y}\left[\min \mu_{\tilde{A}}(x), \mu_{\tilde{B}}(x)\right]; \\ \tilde{C} = \tilde{A} - \tilde{B} \Leftrightarrow \mu_{\tilde{C}}(z) = \max_{z=x-y}\left[\min \mu_{\tilde{A}}(x), \mu_{\tilde{B}}(x)\right]; \\ \tilde{C} = \tilde{A} \otimes \tilde{B} \Leftrightarrow \mu_{\tilde{C}}(z) = \max_{z=x\cdot y}\left[\min \mu_{\tilde{A}}(x), \mu_{\tilde{B}}(x)\right]; \\ \tilde{C} = \tilde{A} : \tilde{B} \Leftrightarrow \mu_{\tilde{C}}(z) = \max_{z=x:y}\left[\min \mu_{\tilde{A}}(x), \mu_{\tilde{B}}(x)\right], \quad y \neq 0. \end{array}\right\} \qquad (1.5)$$

FN of $(L-R)$-type [15] are frequently used to solve problems in the various areas. The following conditions are superimposed on functions L and R:

1. $L(0) = R(0) = 1; L(1) = R(1) = 1;$
2. $L(x)$ and $R(x)$ are nonincreasing functions at $\forall x \in [0,1]$.

FN $\tilde{A}$ with the following membership function

$$\mu_{\tilde{A}}(x) = \begin{cases} L\left\{\dfrac{a_1 - x}{a_L}\right\}, & 0 < \dfrac{a_1 - x}{a_L} \leq 1, \quad a_L > 0; \\ R\left(\dfrac{x - a_2}{a_R}\right), & 0 < \dfrac{x - a_2}{a_R} \leq 1, \quad a_R > 0; \\ 1, & a_1 \leq x \leq a_2; \\ 0, & x < a_1 - a_L \quad \text{or} \quad x > a_2 + a_R. \end{cases} \tag{1.6}$$

is referred to as tolerance $(L-R)$-number.

FN $\tilde{A}$ is symbolically written in the form $\tilde{A} \equiv (a_1, a_2, a_L, a_R)$ [or $\mu_{\tilde{A}}(x) \equiv (a_1, a_2, a_L, a_R)$], where a_1, a_2, a_L, a_R are parameters of tolerance $(L-R)$-number $\tilde{A}$; a segment $[a_1, a_2]$ is a tolerance interval; and a_L, a_R are left and right coefficients of fuzziness, accordingly;

$$L\left(\dfrac{a_1 - x}{a_L}\right) \quad \text{and} \quad R\left(\dfrac{x - a_2}{a_R}\right)$$

are left and right boundaries of membership function of tolerance $(L-R)$-number: with $a_L = 0$

$$L\left(\dfrac{a_1 - x}{a_L}\right) = 0,$$

with $a_R = 0$

$$R\left(\dfrac{x - a_2}{a_R}\right) = 0.$$

The unimodal $\tilde{A}$-$(L-R)$ number has membership function of tolerance $(L-R)$-number under the condition of $a_1 = a_2$. A unimodal $(L-R)$ number is written symbolically as $\tilde{A} \equiv (a_1, a_L, a_R)$. If $\tilde{A} \equiv (a_1, a_2, a_L, a_R)$, $\tilde{B} \equiv (b_1, b_2, b_L, b_R)$, then:

1. $\tilde{A} \oplus \tilde{B} \equiv (a_1 + b_1, a_2 + b_2, a_L + b_L, a_R + b_R);$
2. $\beta\tilde{A} \equiv (\beta a_1, \beta a_2, \beta a_L, \beta a_R)$ on $\beta \geq 0$
3. $\beta\tilde{A} \equiv (\beta a_2, \beta a_1, |\beta|a_R, |\beta|a_L)$ on $\beta < 0$

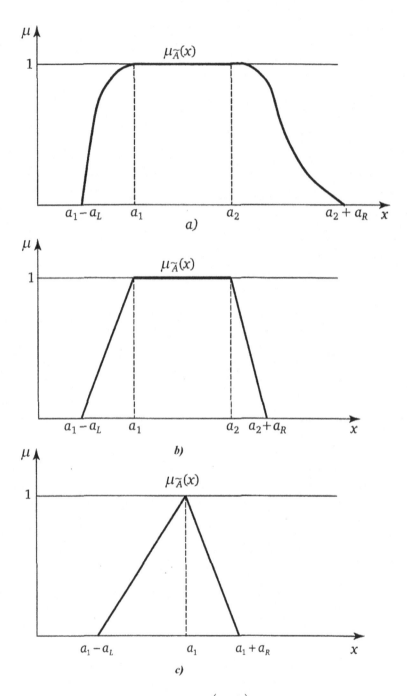

Fig. 1.4 Membership function of FN: **a** — $(L-R)$-type; **b** — T-type; **c** — normal triangular type

By analogy with operation of algebraic product (multiplication), exponentiation operation is defined. If for FN $\tilde{A}$ membership function is $\mu_{\tilde{A}}$, then for FN $\tilde{A}^{\gamma}$ its membership function is $\mu_{\tilde{A}}^{\gamma}$.

Let us give examples of the frequently used FN, or to be exact, membership functions which define these FN (fig. 1.4).

1. Membership function of $(L-R)$-type FN (fig. 1.4.a)

$$L(x)=1-x^2, \quad 0\le x\le 1; \quad R(x)=\begin{cases}1-2x^2, & 0\le x\le \dfrac{1}{2};\\[2mm] 2(x-1)^2, & \dfrac{1}{2}<x\le 1,\end{cases}$$

$$\mu_{\tilde{A}}(x)=\begin{cases}0, & x\le a_1-a_L;\ x>a_2+a_R;\\[2mm] 1-\left(\dfrac{a_1-x}{a_L}\right)^2, & a_1-a_L<x\le a;\\[2mm] 1, & a_1<x\le a_2;\\[2mm] 1-2\left(\dfrac{x-a_2}{a_R}\right)^2, & a_2<x\le a_2+\dfrac{a_R}{2};\\[2mm] 2\left(\dfrac{x-a_2}{a_R}-1\right)^2, & a_2+\dfrac{a_R}{2}<x\le a_2+a_R.\end{cases}$$

2. Membership function of T-type FN (FN of trapezoidal type) $L(x)=R(x)=1-x$, $0\le x\le 1$ (Fig. 1.4.b)

$$\mu_{\tilde{A}}(x)=\begin{cases}1-\dfrac{a_1-x}{a_L}, & 0<\dfrac{a_1-x}{a_L}\le 1,\ a_L>0;\\[2mm] 1-\dfrac{x-a_2}{a_R}, & 0<\dfrac{x-a_2}{a_R}\le 1,\ a_R>0;\\[2mm] 1, & a_1\le x\le a_2;\\[2mm] 0, & x<a_1-a_L \ \text{ or } \ x>a_2+a_R.\end{cases}$$

3. Membership function of a normal triangular number or normal FN of triangular type (Fig. 1.4.c)

$$L(x)=R(x)=1-x, \quad 0\le x\le 1; \quad \mu_{\tilde{A}}(x)=(a_1,a_L,a_R)$$

$$\mu_{\tilde{A}}(x)=\begin{cases}1-\dfrac{a_1-x}{a_L}, & 0<\dfrac{a_1-x}{a_L}\le 1,\ a_L>0;\\[2mm]1-\dfrac{x-a_1}{a_R}, & 0<\dfrac{x-a_1}{a_R}\le 1,\ a_R>0;\\[2mm]1, & x=a_1;\\[2mm]0, & x<a_1-a_L\quad\text{or}\quad x>a_1+a_R.\end{cases}$$

1.4 Linguistic Variables and Semantic Spaces

One of the basic concepts of the fuzzy set theory is the concept of a fuzzy variable [4].

A triple

$$\{X,U,\tilde{A}\}$$

is referred to as a fuzzy variable, where X – the name of the variable; U — area of its definition (universal set); $\tilde{A}$ — the fuzzy set of universal set which describes possible values of the fuzzy variable.

On the basis of concept of a fuzzy variable the concept of a linguistic variable is introduced.

A quintuple

$$\{X,T(X),U,V,S\},$$

is referred to as a linguistic variable, where $T(X)=\{X_i,\ i=\overline{1,m}\}$ — a term-set of a variable X, i.e. set of terms or titles of linguistic values of the variable (each of them is a fuzzy variable with values from universal set U);

V - Syntactic rule generating titles of values of the linguistic variable X;

S - Semantic rule which puts a fuzzy subset of universal set U in conformity to each fuzzy variable with a title from $T(X)$.

Terms X_i are concepts which form a linguistic variable [28]. Membership functions of fuzzy sets $\tilde{X}_i$ describing possible values of fuzzy variables with titles X_i are traditionally referred to as membership functions of concepts X_i, or membership functions of terms X_i. According to one of psycholinguistics principles, namely, to a principle linguistic complementarity [45], membership functions of the same concepts used by different people, do not necessarily coincide.

As an example of the linguistic variable which is formalization of qualitative characteristics, one can see the linguistic variable of "knowledge" shown in Fig. 1.5.

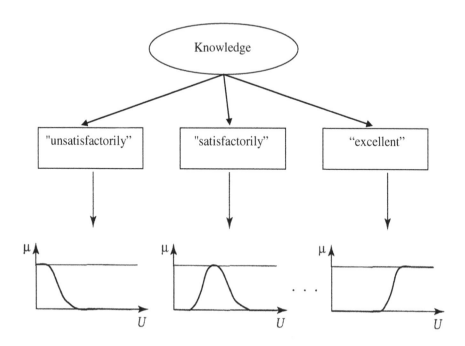

Fig. 1.5 A linguistic variable - "knowledge".

A linguistic variable "RAM space" (memory size) can be considered as an example of the linguistic variable which is formalization of quantitative characteristics.

A linguistic variable with the fixed term-set $\{X,T(X),U,S\}$ is referred to as semantic space.

According to the fuzzy set theory, semantic spaces with a wide spectrum of practical applications can serve as models of expert evaluations of characteristics; these applications include development of expert and intelligence decision making support systems, data analysis and complex process control [4. 35. 37. 47—55].

A semantic space "probability of maintaining solvency by an emitter of the Central bank" can be considered as an example

$$T(X)=\{A,B,C,D,E\},\ U=[0,1],$$

where $A-E$ — probabilities of maintaining solvency (A — very low; B — low; C — mean; D — high; E — very high).

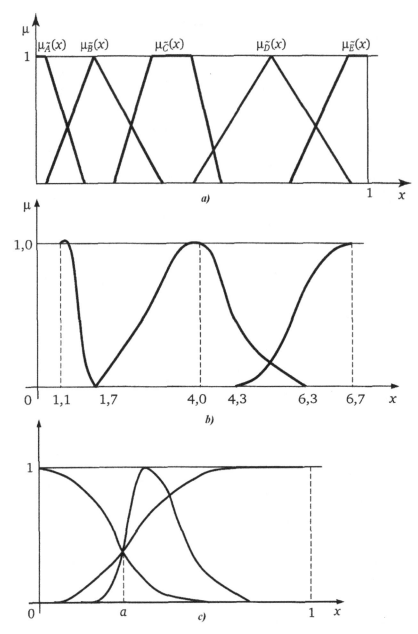

Fig. 1.6 Membership functions of semantic space terms-sets: a — "probability of maintaining solvency by an emitter of the Central bank"; b — "pressure at the high-pressure preheater inlet"; c – "system sensor failure probability"

The choice of universal set is unambiguous here, and term-set membership functions (without generality limitation) are shown in Fig. 1.6. a.

However, not all models based on semantic spaces, possess the properties ensuring successful solutions of practical problems. One such property is completeness of model, which consists in a possibility of describing each element of universal set in linguistic terms of this space [28. 35].

Let us consider a semantic space "steam pressure at inlet" when using a high-pressure preheater, which was constructed in [35],

$$T(X) = \{A, B, C\}, \quad U = [1,1; 6,7],$$

Where A — "low pressure"; B — "pressure close to 4"; C — "high input pressure".

Membership functions of the term-set of this semantic space are shown in Fig. 1.6. b. If at the inlet of a high pressure preheater there is pressure equal 1.7, this value cannot be described by any of linguistic values of the "steam pressure" characteristic. So, we can conclude that the semantic space the membership functions of which are shown in Fig. 1.6. b does not possess the completeness property which means that each element of universal set can be described within the scope of at least one of linguistic terms.

Another model which poorly describes evaluation processes is a semantic space "system sensor failure probability" with terms of "low", "mean", "high" probabilities. Membership functions of this space are shown in Fig. 1.6.c. According to this model, the probability value equal to a is identified with all terms included and consequently it does not have any meaning for further use.

So, a conclusion can be made that the semantic space which membership functions are shown in Fig. 1.6.c, does not possess the property of concepts discriminability which means that each element of universal set cannot be described within the limits of more than two linguistic terms.

It is obvious that to solve practical problems, lack of completeness which characterizes model of expert evaluations of object's properties and lack of discriminability of concepts incorporated into this model, is the essential gap which needs to be bridged.

1.5 Complete Orthogonal Semantic Spaces

Theoretical researches of semantic space properties aimed at improving the adequacy of evaluations expert models and their usefulness for solution of practical problems allow reasonable formulation of requirements to membership functions $\mu_l(x)$; $l = \overline{1, m}$ and their term-sets [28].

1. For each concept X_l the $\hat{U}_l \neq \emptyset$ exists, where $\hat{U}_l = \{x \in U : \mu_l(x) = 1\}$ is a point or a segment.
2. Let $\hat{U}_l = \{x \in U : \mu_l(x) = 1\}$, then $\mu_l(x)$ does not decrease at the left and does not increase to the right of $\hat{U}_l$.
3. $\mu_l(x)$ have no more than two points of discontinuity of the first kind.
4. For every $x \in U$

$$\sum_{l=1}^{m} \mu_i(x) = 1.$$

The semantic spaces with the membership functions being in line with the formulated requirements are called complete orthogonal semantic spaces (COSS).

The requirement 1 means that each concept (term) has at least one standard, which is the typical representative (level of expert gradational confidence of a typical representative's membership to corresponding concept is equal to unity). If there are several standards, all of them are located closely, instead of being scattered over a universal set.

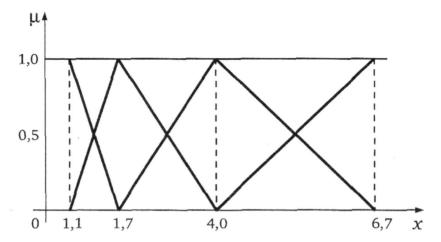

Fig. 1.7 Membership functions of COSS term-set "inlet steam pressure" of the "high pressure preheater"

The requirement 2 says that if units are "close" to each other in universal set, they also are "close" by a membership to a certain concept.

The requirement 3 ensures a possibility of simultaneous processing of fuzzy and precise information from unified positions because regular characteristic functions can be used as membership functions.

The requirement 4 ensures, for each unit from universal set, availability of at least one concept which describes this unit with nonzero grade of membership. Besides, it ensures the discriminability of concepts generating semantic space and excludes the use of synonyms or semantically close terms.

An example is a complete orthogonal semantic space which term-set membership functions are shown in Fig. 1.7.

The formulated requirements to functions of COSS term-set allow defining degree of its fuzziness [28]

$$\zeta = \frac{1}{|U|} \int_U f\big[\mu_{l_1}(x) - \mu_{l_2}(x)\big] dx, \qquad (1.7)$$

Where $\mu_{l_1}(x) = \max_{1 \le l \le m} \mu_l(x);\quad \mu_{l_2}(x) = \max_{\substack{1 \le l \le m \\ l \ne l_1}} \mu_l(x);\quad f(0) = 1;\ f(1) = 0.$

If $L(x) = R(x) = f(x) = 1 - x$, then

$$\zeta = \frac{1}{|U|} \int_U \left[1 - \left[\mu_{l_1}(x) - \mu_{l_2}(x)\right]\right] dx = \frac{|\overline{U}|}{2|U|},$$

where $\overline{U} = U - \bigcup_{l=1}^{m} \widehat{U}_l$.

As COSS can serve as models of expert evaluation of characteristics, then degree of COSS fuzziness is interpreted as mean degree of difficulties of description of real objects and situations by an expert and, besides, as a quantity indicator of quality of the fuzzy information provided by experts.

In [28] it is shown that linear transformation of universal set does not change degree of fuzziness of the relevant COSS.

It is worth mentioning that researches of COSS properties and justification of their employment for various practical problems are under development. The history of these researches repeats history of researches of actually all sections of the fuzzy set theory, where the theoretical component always noticeably lagged behind the practical one.

1.6 Overview of Model-Building Methods for Membership Functions of Fuzzy Sets and Semantic Spaces

Important phase of information processing is information formalization, i.e. its representation in a form allowing application of means of known mathematical theories at subsequent stages of its processing and analysis. For example, if the obtained data are values of some chance quantity, the means of probability theory and the mathematical statistics are applied; and if they are considered as values of some fuzzy variable, means of the fuzzy set theory are applied.

Complexity of formalization of fuzzy information is that applied building methods of membership functions of fuzzy sets and term-sets of semantic spaces are beyond the fuzzy set theory, and therefore adequacy of formalization models cannot be checked up within the scope of its means.

Let us consider known building methods of membership functions of fuzzy sets and term-sets of semantic spaces which are directly set by a table, a formula, and an example [56—59].

Example 1.5. Various representations of a fuzzy set. The fuzzy set $\widetilde{A}$ = {business success of various strategies of a financial structure development} is possible to represent as:

$$x_1 \dots\dots\dots\quad 1\quad 2\quad 3\quad 4$$
$$\mu_{\widetilde{A}}(x_1) \dots\dots\quad 0{,}2\quad 0{,}5\quad 0{,}9\quad 0{,}3$$

The same set can be written in two ways

$$\tilde{A} = \{0,2/1;\ 0,5/2;\ 0,9/3;\ 0,3/4\};$$
$$\tilde{A} = \{0,2/1 + 0,5/2 + 0,9/3 + 0,3/4\}.$$

Example 1.6. Building of fuzzy set on the basis of frequency approach [60—63].

Let us consider fuzzy set $\tilde{A}$ = {the raised demand for cars of the German concern "Volkswagen"}.

Sales volumes of cars of this concern over a certain period are the following: "Bora" — 75, "Passat" — 120, "Golf" — 90, "Jetta" — 60, "Touareg" — 90. It is clear that the "Passat" model was in maximum demand. By introducing a normalization factor 1/120, we'll obtain fuzzy set $\tilde{A}$ = {0. 625 / "Bora", 1 / "Passat", 0.75 / "Golf", 0.5 / "Jetta", 0.75 / "Touareg"}.

In [64], for building of membership function of fuzzy set the probability density function of a continuous chance quantity is evaluated over small volume sample.

Membership function for a certain element of universal set can be defined by ratio of a number of experts considering that this element is typical for fuzzy set, to a number of all experts who are taking part in a survey [35].

Example 1.7. Building of fuzzy set on the basis of an expert group survey. Six independent experts estimated the "correspondence to curricula" characteristic for seven samples of manuals (1—7). Following outcomes were obtained:

Sample No.	Number of experts giving positive (negative) mark
1	2 (4)
2	4 (2)
3	5 0)
4	1 (5)
5	6 (0)
6	3 (3)
7	0 (6)

Thus, this fuzzy set can be written as

$$\left\{ \frac{1}{3}/1;\ \frac{2}{3}/2;\ \frac{5}{6}/3;\ \frac{1}{6}/4;\ 1/5;\ \frac{1}{2}/6;\ 0/7 \right\}.$$

If it is guaranteed that experts are far from random errors, they can be directly interrogated about values of membership function. It is important to remember about distortions of evaluations which are expressed rather often in the subjective tendency to shift them to extremities of an estimating scale [59].

Membership functions can be constructed so that to satisfy the side conditions formulated in advance which can be imposed either to form of obtained information, or to processing procedure. Examples of such side conditions are as follows:

- Membership function shall reflect affinity to a pre-selected standard, and evaluation objects are points in parametrical space [65];
- Membership function shall satisfy the conditions of an interval scale [66];
- Under paired comparison of objects, if one object is estimated as being β –times stronger than the other one, the latter shall be only estimated as being $1/\beta$ – times stronger than the first one [67];
- Restrictions on elements of paired comparison matrix should be introduced when it is not obviously possible to obtain all its values [68], etc.

Example 1.8. Building of a fuzzy set on the basis of paired comparisons of objects [69—70]. An expert estimates the public importance of four innovative projects under a scale presented in table 1.1.

The matrix of paired comparisons is formed as follows: diagonal elements are equal to 1, and for the elements which are symmetric versus a diagonal,

$$a_{ij} = \frac{1}{a_{ji}}, \ i = \overline{1,4}, \ j = \overline{1,4}.$$

Element a_{ij} of a paired comparison matrix is an evaluation of preference of i-th project over j-th by their public importance with its effect on an investor's image taken into account.

The expert composed a matrix of paired comparisons in the following form:

$$A = \begin{pmatrix} 1 & 5 & 6 & 7 \\ \frac{1}{5} & 1 & 4 & 6 \\ \frac{1}{6} & \frac{1}{4} & 1 & 4 \\ \frac{1}{7} & \frac{1}{6} & \frac{1}{4} & 1 \end{pmatrix}.$$

Then, it is necessary to define eigenvalues of the matrix A and to select a maximum eigenvalue. Let us equate to zero a determinant $|A - \lambda E| = 0$:

$$\begin{vmatrix} 1-\lambda & 5 & 6 & 7 \\ \frac{1}{5} & 1-\lambda & 4 & 6 \\ \frac{1}{6} & \frac{1}{4} & 1-\lambda & 4 \\ \frac{1}{7} & \frac{1}{6} & \frac{1}{4} & 1-\lambda \end{vmatrix} = \lambda^4 - 4\lambda^3 - 6{,}914\lambda - 2{,}715 = 0.$$

Table 1.1 A scale of a judgment matrix definition

Evaluation of importance	Qualitative evaluation	Note
1	Identical importance	By the given criterion, projects have an identical rank
3	Weak superiority	Reasons on preference of a project over the other one are flimsy
5	Strong (or essential) superiority	There are relevant proof of the essential superiority of one of the projects
7	Obvious superiority	There are convincing demonstrations in favour of one of the projects
9	Absolute superiority of one of the projects over the other is extremely evident	Preference of one of the projects is well-founded
2. 4. 6. 8	Intermediate values between the adjacent evaluations	Used when the compromise is necessary

Then

$$\lambda_1 = -0{,}362; \quad \lambda_2 = -0{,}140 + 1{,}305i;$$
$$\lambda_3 = -0{,}140 - 1{,}305i; \quad \lambda_4 = 4{,}390.$$

Let us find an eigenvector for maximum eigenvalue $\lambda_{\text{макс}} = 4{,}390$

$$\begin{pmatrix} 1-4{,}390 & 5 & 6 & 7 \\ \dfrac{1}{5} & 1-4{,}390 & 4 & 6 \\ \dfrac{1}{6} & \dfrac{1}{4} & 1-4{,}390 & 4 \\ \dfrac{1}{7} & \dfrac{1}{6} & \dfrac{1}{4} & 1-4{,}390 \end{pmatrix} \begin{pmatrix} \omega_1 \\ \omega_2 \\ \omega_3 \\ \omega_4 \end{pmatrix} = 0.$$

On multiplying the matrixes, we obtain the following set of equations

$$\begin{cases} -3{,}390\omega_1 + 5\omega_2 + 6\omega_3 + 7\omega_4 = 0; \\ 0{,}200\omega_1 - 3{,}390\omega_2 + 4\omega_3 + 6\omega_4 = 0; \\ 0{,}166\omega_1 + 0{,}250\omega_2 - 3{,}390\omega_3 + 4\omega_4 = 0; \\ 0{,}142\omega_1 + 0{,}166\omega_2 + 0{,}250\omega_3 - 3{,}390\omega_4 = 0, \end{cases}$$

which only has a trivial solution.

Let us substitute any of equations of the system with the equation $\omega_1 + \omega_2 + \omega_3 + \omega_4 = 1$, as a result we'll obtain a solution: $\omega_1 = 0{,}619$; $\omega_2 = 0{,}235$; $\omega_3 = 0{,}101$; $\omega_4 = 0{,}045$.

If it is necessary to select a project with the greatest public importance among four projects, the fuzzy set is to be considered

$$\{0{,}619 / №1; \ 0{,}235 / №2; \ 0{,}101 / №3; \ 0{,}045 / №4\}.$$

Then project No. 1 is to be selected because it has the greatest grade of membership to the constructed fuzzy set.

In [71—72] values of membership functions of fuzzy set are defined by an expert group following the rank orderings of objects.

Model-building methods of semantic spaces unlike building methods of separate fuzzy sets are sparse. It is necessary to outline a model-building techniques of linguistic terms of frequency evaluations developed by D.A.Pospelov, I.V.Ezhkova [73—74]. A method shortage is lack of formal algorithm of membership functions building for terms. Ambiguity of building process and of its quality dependence on experience and skills of contributors are consequences of that shortage. The method [73—74] was further developed in methods of S.G. Svarovsky [75] and I.A.Khodashinsky [76].

In these methods connections between membership functions of terms (presence of one maxima and fronts which are smoothly damping to zero are only supposed for membership functions) are not described, moreover, there is no algorithm of deriving of continuous membership functions based on their discrete values. A model-building techniques of the semantic spaces developed by A.V. Skofenko [77] is based on expert evaluations such as "approximately equal to number A" or "lays approximately in the range between A and B". In methods [73—76] connections between membership functions of terms are not described thus not allowing winning independence of quality of the constructed models from skills of researchers.

In [78—79] A.N.Borisov and A.N.Averkin offered parametrical definition of membership functions of the modified terms of semantic spaces on the basis of membership functions of basic terms.

Considering the essential importance of formalization of the information obtained while developing fuzzy models, building methods of COSS membership functions will be considered in Chapter 2.

1.7 Formalization of Fuzzy Conclusions

Let $X_1, X_2, ..., X_n$ be semantic spaces corresponding to universal sets $U_1, U_2, ..., U_n$, accordingly, and terms $\{X_{il}\}$, $i = \overline{1,n}$, $l = \overline{1,m}$, with membership functions $\{\mu_{il}(x)\}$. Let Y be a semantic space with universal set U and terms $\{Y_l\}$, which have membership functions $\{\mu_l(x)\}$.

The system of fuzzy reasonings can be of two types: $X_1, X_2, ..., X_n$ — input information, and Y — output information, or Y is input information, and $X_1, X_2, ..., X_n$ — output information. System of approximate reasonings of first type [80-81] or system of the reference fuzzy logic reasonings which reflect expert experience could be presented as:

$$\tilde{A}^1 = \begin{cases} \tilde{A}_1^1 : \left\langle \begin{array}{l} \text{if } X_{11} \text{ and } X_{21} \text{ and...and } X_{n1} \text{ or } X_{12} \text{ and } X_{21} \text{ and...and } X_{n1} \\ \text{or...or } X_{11} \text{ and } X_{21} \text{ and...and } X_{n2}, \text{ then } Y_1 \end{array} \right\rangle; \\ \tilde{A}_2^1 : \left\langle \begin{array}{l} \text{if } X_{12} \text{ and } X_{22} \text{ and...and } X_{n2} \text{ or } X_{12} \text{ and } X_{23} \text{ and...and } X_{n3} \\ \text{or...or } X_{13} \text{ and } X_{23} \text{ and...and } X_{n2}, \text{ then } Y_2 \end{array} \right\rangle; \\ \hdots\hdots\hdots\hdots\hdots\hdots\hdots\hdots\hdots\hdots \\ \tilde{A}_m^1 : \left\langle \begin{array}{l} \text{if } X_{1m_1} \text{ and } X_{2m_2} \text{ and...and } X_{nm_n} \text{ or } X_{1m_1} \text{ and } X_{2m_2-1} \text{ and...and } X_{nm_n} \\ \text{or...or } X_{1m_1-1} \text{ and } X_{2m_2} \text{ and...and } X_{nm_n-1}, \text{ then } Y_m \end{array} \right\rangle. \end{cases}$$

The second type of this system could be presented as:

$$\widetilde{A}^2 = \begin{cases} \widetilde{A}_1^2 : \left\langle \begin{array}{l} \text{if } Y_1, \text{ then } X_{11} \text{ and } X_{21} \text{ and...and } X_{n1} \text{ or } X_{12} \text{ and } X_{21} \text{ and...and } X_{n1} \\ \text{or...or } X_{11} \text{ and } X_{21} \text{ and...and } X_{n2} \end{array} \right\rangle; \\[2ex] \widetilde{A}_2^2 : \left\langle \begin{array}{l} \text{if } Y_2, \text{ then } X_{12} \text{ and } X_{22} \text{ and...and } X_{n2} \text{ или } X_{12} \text{ and } X_{21} \text{ and...and } X_{n2} \\ \text{or...or } X_{13} \text{ and } X_{22} \text{ and...and } X_{n1} \end{array} \right\rangle; \\[1ex] \cdots\cdots\cdots\cdots\cdots\cdots\cdots\cdots\cdots\cdots\cdots\cdots\cdots\cdots\cdots \\ \widetilde{A}_m^2 : \left\langle \begin{array}{l} \text{if } Y_m, \text{ then} X_{1m_1} \text{ and } X_{2m_2} \text{ and...and } X_{nm_n} \text{ or } X_{1m_1-1} \text{ and } X_{2m_2} \text{ and...and } X_{nm_n} \\ \text{or...or } X_{1m_1} \text{ and } X_{2m_2} \text{ and...and } X_{nm_n-1} \end{array} \right\rangle. \end{cases}$$

Let us consider the block diagram of the fuzzy linguistic governor presented in Fig. 1.8.

The numerical information from an object under control comes to fuzzificator. It transforms the information to a linguistic form. Further, linguistic values come to fuzzy conclusion rule system based on fuzzy expert judgments. As a result, membership function comes to "defuzzificator" where it will be transformed to the numerical information used to control an object.

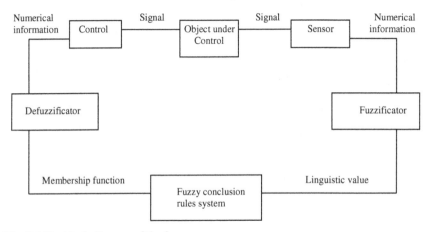

Fig. 1.8 The block diagram of the fuzzy governor

Example 1.9. The fuzzy governor. Let us consider a microcontroller operating as a governor. Its input analogue signal e is limited by a range [-1. 1] and is converted to a digital signal by an analog-to-digital converter with resolution 0.25; the output signal x is formed by the digital-to-analogue converter and has five levels:-1, -0.5, 0, 0.5, 1 [82]. Let us denote with Δe the first error signal difference at the current discrete point of time.

Let us consider semantic space of an analogue signal with terms: A_1 — "high positive", A_2 — "low positive", A_3 — "zero", A_4 — "low negative", A_5 — "high negative" and semantic space "the first error signal difference at the current discrete point of time" with terms: B_1 — "high positive", B_2 — "low positive", B_3 — "zero", B_4 — "low negative", B_5 — "high negative". Membership functions of terms of those two spaces are assumed identical, values of membership functions A_i, $i = \overline{1,5}$ are presented in Table 1.2.

Table 1.2 Membership functions of terms e

$\mu_{\tilde{A}_i}$	−1	−0.75	−0.5	−0.25	0	0.25	0.5	0.75	1
$\mu_{\tilde{A}_1}$	0	0	0	0	0	0	0.3	0.7	1
$\mu_{\tilde{A}_2}$	0	0	0	0	0.3	0.7	1	0.7	0.3
$\mu_{\tilde{A}_3}$	0	0	0.3	0.7	1	0.7	0.3	0	0
$\mu_{\tilde{A}_4}$	0.3	0.7	1	0.7	0.3	0	0	0	0
$\mu_{\tilde{A}_5}$	1	0.7	0.3	0	0	0	0	0	0

Let us assume that functioning of the governor is defined by following relations;

e	Δe	x
A_1	B_1	−1
A_2	B_2	−0.5
A_3	B_3	0
A_4	B_4	0.5
A_5	B_5	1

Certainly, collections of relations can be different. Following the formulated relations, let us define an output signal of the governor under $e = 0,25$; $\Delta e = 0,5$. Let us find values of membership functions of terms in points $e = 0,25$; $\Delta e = 0,5$, using Table 1.2. We'll denote $\alpha_i = \min[A_i(e), B_i(\Delta e)]$ then:

$$\alpha_1 = \min(0;0,3) = 0; \ x_1 = -1;$$
$$\alpha_2 = \min(0,7;1) = 0,7; \ x_2 = -0,5;$$
$$\alpha_3 = \min(0,7;0,3) = 0,3; \ x_3 = 0;$$
$$\alpha_4 = \min(0;0) = 0; \ x_4 = 0,5;$$
$$\alpha_5 = \min(0;0) = 0; \ x_5 = 1;$$

Output signal of the governor

$$x = x_k, \ \text{if} \ \alpha_k = \max \alpha_i; \ i = \overline{1,5}; \ k = \overline{1,5}.$$

Maximum value of α_i is equal to 0.7, therefore the output signal of the governor is equal to -0.5.

There are some algorithms of a fuzzy logic conclusion. Let us consider most known of them, using system of two fuzzy conclusion rules:

$$\text{if } X \text{ is } \tilde{A}_1 \text{ and } Y \text{ is } \tilde{B}_1, \text{ then } Z \text{ is } \tilde{C}_1;$$

$$\text{if } X \text{ is } \tilde{A}_2 \text{ and } Y \text{ is } \tilde{B}_2, \text{ then } Z \text{ is } \tilde{C}_2.$$

and having denoted membership functions of linguistic values of variables X, Y, Z through $\mu_{\tilde{A}_i}$, $\mu_{\tilde{B}_i}$, $\mu_{\tilde{C}_i}$, $i = \overline{1,2}$ accordingly.

Algorithm of Mamdani

1. Fuzzification. Membership functions $\mu_{\tilde{A}_i}$, $\mu_{\tilde{B}_i}$, $i = \overline{1,2}$ are applied to physical (real) values x_0, y_0 of variables X, Y.

2. Fuzzy conclusion

$$\alpha_1 = \min\left[\mu_{\tilde{A}_1}(x_0), \mu_{\tilde{B}_1}(y_0)\right];$$

$$\alpha_2 = \min\left[\mu_{\tilde{A}_2}(x_0), \mu_{\tilde{B}_2}(y_0)\right];$$

Definition of the truncated membership functions of variable Z:

$$\mu_{\tilde{C}_1^1} = \min\left[\alpha_1, \mu_{\tilde{C}_1}(z)\right];$$

$$\mu_{\tilde{C}_2^1} = \min\left[\alpha_2, \mu_{\tilde{C}_2}(z)\right]$$

3. A composition (union of the truncated membership functions)

$$\mu_{\Sigma}(z) = \max\left[\mu_{\tilde{C}_1^1}(z), \mu_{\tilde{C}_2^1}(z)\right];$$

4. Defuzzification:
For a continuous case

$$z_0 = \frac{\int\limits_U z\mu_{\Sigma}(z)dz}{\int\limits_U \mu_{\Sigma}(z)dz};$$

For a discrete case

$$z_0 = \frac{\sum\limits_i z_i\mu_{\Sigma}(z_i)}{\sum\limits_i \mu_{\Sigma}(z_i)}.$$

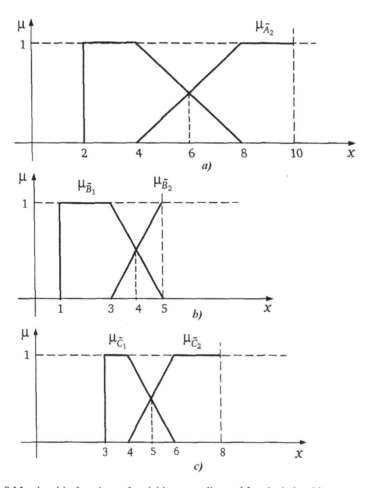

Fig. 1.9 Membership functions of variables according to Mamdani algorithm:

$$\mathbf{a} - X(\tilde{A}_1, \tilde{A}_2); \ \mathbf{b} - Y(\tilde{B}_1, \tilde{B}_2); \ \mathbf{c} - Z(\tilde{C}_1, \tilde{C}_2)$$

Example 1.10. Application of Mamdani algorithm. Let us assume that ranges of values of variables X, Y, Z are segments [2; 10], [1; 5], [3; 8], accordingly. Linguistic values of these variables ($X - \tilde{A}_1, \tilde{A}_2$, $Y - \tilde{B}_1, \tilde{B}_2$, $Z - \tilde{C}_1, \tilde{C}_2$) have the membership functions shown in Fig. 1.9.

Let $x_0 = 5$; $y_0 = 4,8$, then $\mu_{\tilde{A}_1}(5) = 0,75$; $\mu_{\tilde{A}_2}(5) = 0,25$; $\mu_{\tilde{B}_1}(4,8) = 0,1$; $\mu_{\tilde{B}_2}(4,8) = 0,9$.

Then

$$\alpha_1 = \min\left[\mu_{\tilde{A}_1}(x_0), \mu_{\tilde{B}_1}(y_0)\right] = \min(0{,}75;\ 0{,}1) = 0{,}1;$$

$$\alpha_2 = \min\left[\mu_{\tilde{A}_2}(x_0), \mu_{\tilde{B}_2}(y_0)\right] = \min(0{,}25;\ 0{,}9) = 0{,}25.$$

The composition of the truncated membership functions is presented in Fig. 1.10. As a result we'll obtain:

$$z_0 = \frac{\int_3^{4,2} 0{,}1x\,dx + \int_{4,2}^{4,5}(0{,}5x-2)x\,dx + \int_{4,5}^{8} 0{,}25x\,dx}{\int_3^{4,2} 0{,}1\,dx + \int_{4,2}^{4,5}(0{,}5x-2)\,dx + \int_{4,5}^{8} 0{,}25\,dx} =$$

$$= \frac{0{,}05(4{,}2^2-3^2)+0{,}167(4{,}5^3-4{,}2^3)-(4{,}5^2-4{,}2^2)+}{0{,}1(4{,}2-3)+0{,}25(4{,}5^2-4{,}2^2)-2(4{,}5-4{,}2)+} \rightarrow$$

$$\rightarrow \frac{+0{,}125(8^2-4{,}5^2)}{+0{,}25(8-4{,}5)} = 5{,}86.$$

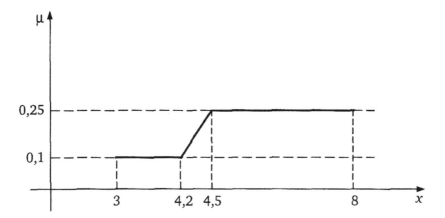

Fig. 1.10 A composition of the truncated membership functions

Algorithm of Tsukamoto

It is assumed that membership functions $\mu_{\tilde{C}_1}(z)$, $\mu_{\tilde{C}_2}(z)$ are monotone.

1. Fuzzification. The same as in Mamdani algorithm.
2. Fuzzy conclusion (the same as in Mamdani algorithm)

$$\alpha_1 = \min\left[\mu_{\tilde{A}_1}(x_0), \mu_{\tilde{B}_1}(y_0)\right];$$

$$\alpha_2 = \min\left[\mu_{\tilde{A}_2}(x_0), \mu_{\tilde{B}_2}(y_0)\right]$$

Precise values z_1, z_2 are derived from the equations

$$\alpha_1 = \mu_{\tilde{C}_1}(z_1);$$

$$\alpha_2 = \mu_{\tilde{C}_2}(z_2).$$

3. Definition of a precise value of a conclusion variable

$$z_0 = \frac{\alpha_1 z_1 + \alpha_2 z_2}{\alpha_1 + \alpha_2}.$$

Example 1.11. Application of Tsukamoto algorithm. As in the example 1.10, let us assume that ranges of values of variables X, Y, Z are segments [2.10], [1.5], [3.8], accordingly. Linguistic values of variables (X — $\tilde{A}_1, \tilde{A}_2$, Y — $\tilde{B}_1, \tilde{B}_2$, Z — $\tilde{C}_1, \tilde{C}_2$) have the membership functions shown in Fig. 1.11.

Let $x_0 = 5$; $y_0 = 4{,}8$, then $\mu_{\tilde{A}_1}(5) = 0{,}75$; $\mu_{\tilde{A}_2}(5) = 0{,}25$; $\mu_{\tilde{B}_1}(4{,}8) = 0{,}1$; $\mu_{\tilde{B}_2}(4{,}8) = 0{,}9$.

$$\alpha_1 = \min\left[\mu_{\tilde{A}_1}(x_0); \mu_{\tilde{B}_1}(y_0)\right] = \min(0{,}75; 0{,}1) = 0{,}1;$$

$$\alpha_2 = \min\left[\mu_{\tilde{A}_2}(x_0); \mu_{\tilde{B}_2}(y_0)\right] = \min(0{,}25; 0{,}9) = 0{,}25.$$

Let us find z_1, z_2 from the equations:

$$0{,}1 = \mu_{\tilde{C}_1} = 1 - \frac{z_1 - 3}{5};$$

$$0{,}25 = \mu_{\tilde{C}_2} = 1 + \frac{z_2 - 8}{5}.$$

Then $z_1 = 3{,}5$; $z_2 = 4{,}25$, and the defuzzification gives:

$$z_0 = \frac{3{,}5 \cdot 0{,}1 + 4{,}25 \cdot 0{,}25}{0{,}1 + 0{,}25} = 4{,}036.$$

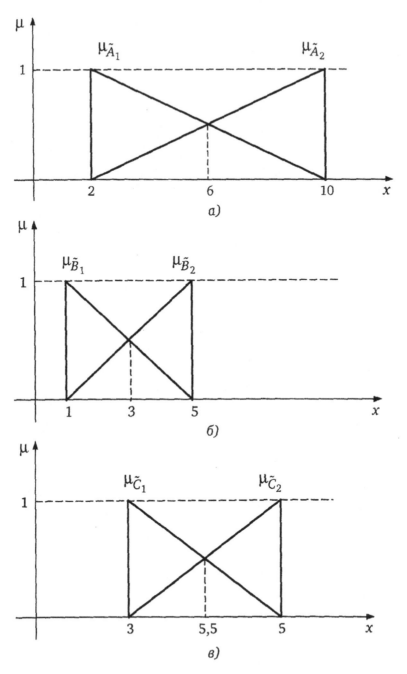

Fig. 1.11 Membership functions of variables according to Tsukamoto algorithm (a—c are the same as in Fig. 1.9)

Sugeno Algorithm

For this algorithm the following conclusion rules are used:

$$\text{if } X \text{ is } \tilde{A}_1 \text{ and } Y \text{ is } \tilde{B}_1 \text{ then } z = a_1 x + b_1 y \text{ ;}$$

$$\text{if } X \text{ is } \tilde{A}_2 \text{ and } Y \text{ is } \tilde{B}_2 \text{ then } z = a_2 x + b_2 y \text{ .}$$

1. Fuzzification is the same as in Mamdani algorithm.
2. A fuzzy conclusion is the same as in Mamdani algorithm.

$$\alpha_1 = \min\left[\mu_{\tilde{A}_1}(x_0), \mu_{\tilde{B}_1}(y_0)\right]$$

$$\alpha_2 = \min\left[\mu_{\tilde{A}_2}(x_0), \mu_{\tilde{B}_2}(y_0)\right]$$

Individual outputs of the rules:

$$z_1 = a_1 x_0 + b_1 y_0;$$

$$z_2 = a_2 x_0 + b_2 y_0.$$

3. Definition of precise value of a conclusion variable (the same as in Tsukamoto algorithm)

$$z_0 = \frac{\alpha_1 z_1 + \alpha_2 z_2}{\alpha_1 + \alpha_2}.$$

Larsen Algorithm

1. Fuzzification is the same as in Mamdani algorithm.
2. Fuzzy conclusion is the same as in Mamdani algorithm.

$$\alpha_1 = \min\left[\mu_{\tilde{A}_1}(x_0), \mu_{\tilde{B}_1}(y_0)\right]$$

$$\alpha_2 = \min\left[\mu_{\tilde{A}_2}(x_0), \mu_{\tilde{B}_2}(y_0)\right]$$

Definition of particular fuzzy sets:

$$\mu_{\tilde{C}_1^1} = \alpha_1 \mu_{\tilde{C}_1}(z);$$

$$\mu_{\tilde{C}_2^1} = \alpha_2 \mu_{\tilde{C}_2}(z).$$

3. Composition. Union of truncated membership functions (the same as in Mamdani algorithm)

$$\mu_\Sigma(z) = \max\left[\mu_{\tilde{C}_1^1}(z), \mu_{\tilde{C}_2^1}(z)\right]$$

4. Defuzzification (the same as in Mamdani algorithm):
for a continuous case

$$z_0 = \frac{\int_U z \mu_\Sigma(z) dz}{\int_U \mu_\Sigma(z) dz};$$

For a discrete case

$$z_0 = \frac{\sum_i z_i \mu_\Sigma(z_i)}{\sum_i \mu_\Sigma(z_i)}.$$

Simplified Algorithm of a Fuzzy Conclusion

In this algorithm the following derivation rules are used:

$$\text{if } x \text{ is } \tilde{A}_1, \text{ and } y \text{ have } \tilde{B}_1 \text{ then } z = c_1;$$

$$\text{if } x \text{ is } \tilde{A}_2 \text{ and } y \text{ have } \tilde{B}_2 \text{ then } z = c_2.$$

where c_1, c_2 — some precise (regular) numbers.

1. Fuzzification is the same as in Mamdani algorithm.
2. Fuzzy conclusion is the same as in Mamdani algorithm.

$$\alpha_1 = \min\left[\mu_{\tilde{A}_1}(x_0), \mu_{\tilde{B}_1}(y_0)\right]$$

$$\alpha_2 = \min\left[\mu_{\tilde{A}_2}(x_0), \mu_{\tilde{B}_2}(y_0)\right]$$

3. Definition of precise value of a conclusion variable

$$z_0 = \frac{\alpha_1 c_1 + \alpha_2 c_2}{\alpha_1 + \alpha_2}.$$

Methods of handling of the fuzzy information presented as a population of approximate judgments are developed well enough. Semantic spaces are usually used as $X_1, X_2, ..., X_n$ and Y. However use of subjective representations of membership functions values of semantic space terms to formalize this information leads to necessity in intensified development aimed to improve adequacy of developed fuzzy models. New relevant outcomes are discussed in Chapter 2.

1.8 Methods of Defuzzification

Defuzzification of a fuzzy number is referred to as its mapping to a point or a segment of the real line. Methods of defuzzification of fuzzy numbers play an essential role for issues of processing and analysis of fuzzy information. There are different methods of defuzzification of fuzzy numbers, in particular, a gravity method, a minimax method, a method of a maximum of membership function, a method of the weighed point etc. [40. 46. 83—93] which have the certain merits and demerits described in [87]. Most frequently applied of these methods is the gravity method, according to which pointwise value defined under the formula:

$$E = \frac{\int\limits_0^1 x\mu(x)dx}{\int\limits_0^1 \mu(x)dx}.$$ (1.8)

is assigned to fuzzy number with membership function $\mu(x)$.

If $L(x)$, and $R(x)$ in (1.6) are linear functions, and $\mu(x) \equiv (a_1, a_2, a_L, a_R)$, then (1-8) looks like:

$$E = \frac{(a_2)^2 - (a_1)^2 + a_1 a_L + a_2 a_R + \frac{1}{3}\left[(a_R)^2 - (a_L)^2\right]}{2(a_2 - a_1) + a_L + a_R}.$$ (1.9)

The method of the maximum of membership function is applied to unimodal fuzzy numbers; the abscissa of a point of a maximum of its membership function is to be taken as an integral index of fuzzy number. In the minimax method a minimum value of abscissas of membership function's maximum points is to be selected as an integral index of fuzzy number. The essence of the method of the weighed point [87] is that generation of an integral index of a normal triangular number is carried out with accounting for weighs of its α-level sets. If the normal triangular number $\tilde{B}$ with membership function $\mu_{\tilde{B}}(x)$ looks like $\mu_{\tilde{B}}(x) \equiv (b, b_L, b_R)$, the corresponding weighed point is defined by the formula:

$$B = \int\limits_0^1 \frac{1}{2}\left[b - (1-\alpha)b_L + b + (1-\alpha)b_R\right]p(\alpha)d\alpha.$$

Function $p(\alpha)$ at $0 \leq \alpha \leq 1$ is referred to as α-levels' importance distribution density function [87]. This function shall be nonnegative and satisfy a normalization condition:

$$\int\limits_0^1 p(\alpha)d\alpha = 1.$$

A weighed point for a normal triangular number with membership function $\mu_{\tilde{B}}(x) \equiv (b, b_L, b_R)$ and $p(\alpha) = 2\alpha$ is defined in [84] by the formula

$$B = \int_0^1 [b - (1-\alpha)b_L + b + (1-\alpha)b_R]\alpha d\alpha = b + \frac{1}{6}(b_R - b_L).$$

In [93] definition of the weighed point is spread to tolerance $(L-R)$-numbers; and the new method of defuzzification of fuzzy numbers is proposed which assigns to them not a number, but a certain numerical segment called the weighed one.

According to [93], weighed segment of a tolerance $(L-R)$-number $A \equiv (a_1, a_2, a_L, a_R)$ is the segment $[A_1, A_2]$:

$$A_1 = a_1 - la_L; \quad A_2 = a_2 + ra_R; \quad l = \int_0^1 L^{-1}(\alpha)\alpha d\alpha; \quad r = \int_0^1 R^{-1}(\alpha)\alpha d\alpha. \quad (1.10)$$

Before the weighed segment being defined, the known methods of defuzzification of fuzzy numbers have allowed definition only pointwise aggregating indexes for them. Thereby, for example, they made unimodal numbers with different fuzziness coefficients or tolerance numbers indiscernible, i.e. their informational singularities were lost, which were necessary to be outlined. In more details, the method of defuzzification of fuzzy numbers based on the weighed segments is discussed in Chapter 6.

1.9 Fuzzy Relations and Features of Fuzzy Cluster Analysis

One of the basic concepts of the fuzzy cluster analysis is the concept of the fuzzy relation. Fuzzy relations play an essential role in the problems with the solutions rested upon methods of the fuzzy set theory, on traditional methods and the theory of clear ratios [94—99]. As a rule, the means of the theory of clear relations is used in qualitative analysis of correlations between objects of investigated system when they are of dichotomizing nature and, consequently, can be interpreted in terms "there is no connection", "there is a connection", or when methods of quantitative analysis of correlations are inapplicable for any reasons, and correlations are derived to a dichotomizing form synthetically. For example, when connection between objects takes values from a rank scale, then selection of connection force threshold allows this connection to be transformed to a demanded kind. However, despite the similar approach allows carrying out the qualitative analysis of systems, it leads to loss of the information on connection force between objects; therefore it is necessary to make evaluations at different connection force thresholds. Methods of data analysis based on the theory of fuzzy ratios do not suffer this shortage; they allow carrying out the qualitative analysis of systems taking into account distinctions in connection force between objects of the system [15].

Ordinary (definite) n-ary relation R between sets $X_1, X_2, ..., X_n$ is referred to as a subset of Cartesian product:

$$R \subseteq X_1 \times X_2 \times .... \times X_n.$$

Fuzzy n-ary relation R between sets $X_1, X_2, \ldots, X_n$ is referred to as such fuzzy set R when $\forall (x_1, x_2, \ldots, x_n) \in X_1 \times X_2 \times \ldots \times X_n$; $\mu_R(x_1, x_2, \ldots, x_n) \in [0, 1]$, where $X_1 = \{x_1\}$; $X_2 = \{x_2\}$, ..., $X_n = \{x_n\}$ are ordinary sets.

Fuzzy binary relation R between sets X, Y is referred to as such fuzzy set R when $\forall (x, y) \in X \times Y$; $\mu_R(x, y) \in [0, 1]$, where $X = \{x\}$; $Y = \{y\}$ are ordinary sets.

If sets $X = \{x_1, x_2, \ldots, x_n\}$; $Y = \{y_1, y_2, \ldots, y_m\}$ are finite, the fuzzy binary relation R can be specified by means of a matrix, which rows and columns are put in correspondence with elements of sets X, Y; and an element $\mu_R(x_i, y_j)$ is put as cross-section the of i-th row and j-th column. Thus

$$
R = \begin{pmatrix}
\mu_R(x_1, y_1) & \mu_R(x_1, y_2) & \cdots & \mu_R(x_1, y_m) \\
\mu_R(x_2, y_1) & \mu_R(x_2, y_2) & \cdots & \mu_R(x_2, y_m) \\
\cdots & \cdots & \cdots & \cdots \\
\mu_R(x_n, y_1) & \mu_R(x_n, y_2) & & \mu_R(x_n, y_m)
\end{pmatrix}.
$$

Fuzzy binary relation R over set X is referred to as such fuzzy set R, that $\forall (x, y) \in X \times Y$; $\mu_R(x, y) \in [0, 1]$.

Let R_1, R_2 are fuzzy binary relations between sets X, Y and Y, Z, accordingly.

Composition of fuzzy relations R_1, R_2 is referred to as such fuzzy set $R_1 \circ R_2$ when for $\forall x \in X$, $\forall y \in Y$, $\forall z \in Z$

$$
\mu_{R_1 \circ R_2}(x, z) = \vee_y \left[\mu_{R_1}(x, y) \wedge \mu_{R_2}(y, z) \right],
$$

where $\wedge$ and $\vee$ are operators of triangular norm and conorm class, accordingly.

For example, ($\max - \min$)-composition $R_1 \circ R_2$ is defined by the expression

$$
\mu_{R_1 \circ R_2}(x, z) = \max_y \left[\min \{ \mu_{R_1}(x, y), \mu_{R_2}(y, z) \} \right],
$$

where $x \in X$; $y \in Y$; $x \in X$.

The fuzzy binary relation R is referred to as reflective, if $\mu_R(x, x) = 1$; $\forall x \in X$, and symmetric, if $\mu_R(x, y) = \mu_R(y, x)$; $\forall x, y \in X$.

One of the important properties of fuzzy binary relations is that they can be presented as a population of ordinary binary relations ordered by inclusion and representing a hierarchical population of relations [15]. Expansion of fuzzy binary relations to a population of ordinary binary relations is based on the concept of α-level of a fuzzy binary relation which is represented in the form

$$
R_\alpha = \{ (x, y) \in X \times Y : \mu_R(x, y) \geq \alpha \}.
$$

Theorem of Decomposition [40]. Any fuzzy set $\tilde{A}$ with membership function $\mu_{\tilde{A}}(x)$ can be decomposed on sets of α-level

$$\tilde{A} = \bigcup_{\alpha \in [0,1]} \alpha A_\alpha$$

(1.11)

or

$$\mu_{\tilde{A}}(x) = \bigcup_{\alpha \in [0,1]} \left[\alpha \mu_{A_\alpha}(x) \right],$$

where $\mu_{A_\alpha}(x) = \begin{cases} 1, & \mu_A(x) \geq \alpha; \\ 0, & \mu_A(x) < \alpha. \end{cases}$

According to (1.11), the fuzzy binary relation R can be presented as

$$R = \bigcup_{\alpha \in [0,1]} \alpha R_\alpha$$

or

$$\mu_R(x) = \bigcup_{\alpha \in [0,1]} \left[\alpha \mu_{R_\alpha}(x) \right],$$

where $\mu_{R_\alpha}(x) = \begin{cases} 1, & \mu_R(x) \geq \alpha; \\ 0, & \mu_R(x) < \alpha. \end{cases}$

A reflective symmetric fuzzy binary relation is referred to as fuzzy binary relation of similarity.

Fuzzy binary relation R is referred to as transitive, if $\mu_R(x, z) \geq \mu_R(x, y) \wedge \mu_R(y, z) \forall x, y, z \in X$.

A transitive fuzzy binary relation of similarity is referred to as fuzzy binary relation of conformity.

In actual practice the transitivity requirement is often difficult to meet. In order to use expert survey for the purpose of constructing the similarity relation, the transitive answers shall be demanded from these experts. Numerous practical outcomes [16] are quite opposite: real outcomes of expert surveys are often intransitive. However, in applications of fuzzy relations the transitive ones are of great importance, because they possess many convenient properties and define some correct structure of a set which they are set for. For example, if relation R over a set X characterizes similarity between objects, then transitivity of such relation (the similarity relation) ensures a possibility of a partition of set X on disjoint similarity classes (clusters).

Let R be similarity relation. Then according to the theorem of decomposition for relations of similarity [15]

$$R = \bigcup_\alpha (\alpha \times R_\alpha),$$

(1.12)

where $(\alpha \times R_\alpha)(x, y) = \begin{cases} \alpha, & \text{if } R_\alpha(x, y) = 1; \\ 0, & \text{otherwise.} \end{cases}$

Thus, to solve problems of fuzzy cluster analysis, possibility to transform an initial intransitive relation to a transitive one is of great interest. Such transformation is ensured with operation of transitive closure which was first considered in [100—101].

The relation $\hat{R} = R \cup R^2 \cup R^3 \cup R^k \cup ...$, where relation R^k is defined recursively as

$$R^2 = R \circ R; \quad R^k = R \circ R^{k-1}, \quad k = 3, 4,...$$

is referred to as transitive closure of a fuzzy binary relation.

Transitive closure $\hat{R}$ of any fuzzy relation R is transitive and is the least transitive relation including R [15]. In [28] it is proved that a fuzzy relation is transitive in the only case when $R = \hat{R}$, and if the set X contains n elements, then

$$\hat{R} = R \cup R^2 \cup R^3 \cup R^n.$$

Besides, if fuzzy relation R is reflective, then $\hat{R} = R^{n-1}$.

Application of fuzzy relations in the cluster analysis is discussed in [102—103]. In [100] clusterization procedure based on transitive closure of an initial relation of the similarity resulted from the expert survey, is offered. Experts compared similarity between portraits of different people, and on the basis of paired comparison the similarity matrix was built. Various methods of fuzzy clusterization and their connections with traditional methods of the cluster analysis and their practical applications are discussed in [104—120].

Let us suppose that some experts estimate expression of some characteristics at a population of objects. Having in mind that each of these experts has his/her individual evaluation criterion, we cannot exclude that the information obtained from one expert differs from the information obtained from another. As models of an expert evaluation of some characteristic are constructed rested upon this information, it is obvious that the models based on information obtained from various experts will differ.

Thus, if within the characteristic limits some expert evaluation models can be constructed; there is a necessity of their comparative analysis versus the subsequent building of the generalized model. The methods developed by authors and allowing to carry out similar researches are discussed in Chapter 3.

Chapter 2

Methods of Expert Information Formalization Based on Complete Orthogonal Semantic Spaces

2.1 Fuzzy Numbers Used for Formalization of Linguistic Values of Characteristics

Let us consider tolerance and unimodal $(L-R)$-numbers with membership functions

$$\mu_{\tilde{A}}(x) = \begin{cases} L\left(\dfrac{a_1 - x}{a_L}\right), & 0 \le \dfrac{a_1 - x}{a_L} \le 1;\ a_L > 0; \\[2mm] R\left(\dfrac{x - a_2}{a_R}\right), & 0 \le \dfrac{x - a_2}{a_R} \le 1;\ a_R > 0; \\[2mm] 1, & \dfrac{a_1 - x}{a_L} < 0 \cap \dfrac{x - a_2}{a_R} < 0; \\[2mm] 0, & \dfrac{a_1 - x}{a_L} > 1 \cup \dfrac{x - a_2}{a_R} > 1 \end{cases}$$

and following conditions for functions L and R:

1. $L(0) = R(0) = 1;\ L(1) = R(1) = 0;$
2. $L(x)$ and $R(x)$ are monotonically decreasing functions over set $[0,1]$.

Let us denote Λ for a population of all tolerance and unimodal numbers with conditions 1 and 2.

Let us call elements of the population Λ as Λ-numbers which are in turn subdivided into Λ-tolerance and Λ-unimodal numbers.

As L and R are monotonically decreasing functions, the set of α-level of Λ-tolerance number $\tilde{A} \equiv (a_1, a_2, a_L, a_R)$ will look like:

$$A_\alpha \{x \in R : \mu_{\tilde{A}}(x) \ge \alpha\} = \left[A_\alpha^1, A_\alpha^2\right] =$$

$$= \left[a_1 - L^{-1}(\alpha)a_L;\ a_2 + R^{-1}(\alpha)a_R\right],\quad \alpha \in [0,1]. \tag{2.1}$$

O.M. Poleshchuk and E.G. Komarov: Expert Fuzzy Info. Processing, STUDFUZZ 268, pp. 43–86.
springerlink.com   © Springer-Verlag Berlin Heidelberg 2011

Let us consider arithmetical operations for $\wedge$-tolerance numbers $\tilde{A} \equiv \left(a_1, a_2, a_{L_1}, a_{R_1}\right)$, $\tilde{B} \equiv \left(b_1, b_2, b_{L_2}, b_{R_2}\right)$.

The Proposition 2.1. [121] Sum of $\wedge$-tolerance numbers is a $\wedge$-tolerance number.

The poof. By definition of the expanded union operation for tolerance $(L-R)$-numbers $\tilde{A}, \tilde{B}$ with membership functions $\mu_{\tilde{A}}(x)$ and $\mu_{\tilde{B}}(x)$, accordingly, we'll obtain tolerance $(L-R)$-number $\tilde{A} \oplus \tilde{B}$ with membership function

$$\mu(z) = \max_{z=x+y}\left[\min \mu_{\tilde{A}}(x), \mu_{\tilde{B}}(y)\right],$$

which is symbolically noted by parameters $\tilde{A} \oplus \tilde{B} \equiv \left(a_1 + b_1, a_2 + b_2, a_{L_1} + b_{L_2}, a_{R_1} + b_{R_2}\right)$. Let us show that the function which is the left boundary of membership function of this number increases monotonically. Let us write out two sets of α-level of number $\tilde{A} + \tilde{B}$, $\alpha_2 > \alpha_1$ according to (2.1):

$$(A+B)_{\alpha_1} = \left[a_1 + b_1 - L_1^{-1}(\alpha_1)a_{L_1} - L_2^{-1}(\alpha_1)b_{L_2} ; \; a_2 + b_2 + R_1^{-1}(\alpha_1)a_{R_1} + R_2^{-1}(\alpha_1)b_{R_2}\right]$$

$$(A+B)_{\alpha_2} = \left[a_1 + b_1 - L_1^{-1}(\alpha_1)a_{L_1} - L_2^{-1}(\alpha_1)b_{L_2} ; \; a_2 + b_2 + R_1^{-1}(\alpha_1)a_{R_1} + R_2^{-1}(\alpha_1)b_{R_2}\right]$$

As functions L_1, L_2, R_1, R_2 are monotonically decreasing, then at $\alpha_2 > \alpha_1$

$$a_1 + b_1 - L_1^{-1}(\alpha_2)a_{L_1} - L_2^{-1}(\alpha_2)b_{L_2} > a_1 + b_1 - L_1^{-1}(\alpha_1)a_{L_1} - L_2^{-1}(\alpha_1)b_{L_2} ;$$

$$a_2 + b_2 + R_1^{-1}(\alpha_2)a_{R_1} + R_2^{-1}(\alpha_2)b_{R_2} < a_2 + b_2 + R_1^{-1}(\alpha_1)a_{R_1} + R_2^{-1}(\alpha_1)b_{R_2} .$$

Hence, the left boundary of membership function $\tilde{A} \oplus \tilde{B}$ monotonically increases, and the right boundary monotonically decreases, and $\tilde{A} \oplus \tilde{B}$ belongs to $\wedge$. The proposition 2.1 is proved.

Similarly, it is possible to show that sum of $\wedge$-unimodal numbers is a $\wedge$-unimodal number, and the sum of $\wedge$-tolerance and $\wedge$-unimodal numbers is $\wedge$-tolerance number. If $L_1 = L_2 = L$, and $R_1 = R_2 = R$, then sum of these numbers is $(L-R)$-number and belongs to $\wedge$.

The Proposition 2.2. [121] Product (result of multiplication) of $\wedge$-tolerance numbers is a $\wedge$-tolerance number.

The proof. Let us consider a fuzzy number which is the product of $\wedge$-tolerance numbers $\tilde{A} \equiv \left(a_1, a_2, a_{L_1}, a_{R_1}\right)$, $\tilde{B} \equiv \left(b_1, b_2, b_{L_2}, b_{R_2}\right)$, and let us denote it through $\tilde{D} = \tilde{A} \otimes \tilde{B}$. Let write out sets of α-level of numbers $\tilde{A}$ and $\tilde{B}$ according to (2.1)

$$A_\alpha = \left[A_\alpha^1, A_\alpha^2\right] = \left[a_1 - L_1^{-1}(\alpha)a_{L_1}, a_2 + R_1^{-1}(\alpha)a_{R_1}\right]$$

$$B_\alpha = \left[B_\alpha^1, B_\alpha^2\right] = \left[b_1 - L_2^{-1}(\alpha)b_{L_2}, b_2 + R_2^{-1}(\alpha)b_{R_2}\right]$$

The set of α-level of $\tilde{D}$ looks like

$$D_\alpha = \left[\min\left(A_\alpha^1 B_\alpha^1, A_\alpha^1 B_\alpha^2, A_\alpha^2 B_\alpha^1, A_\alpha^2 B_\alpha^2\right), \; \max\left(A_\alpha^1 B_\alpha^1, A_\alpha^1 B_\alpha^2, A_\alpha^2 B_\alpha^1, A_\alpha^2 B_\alpha^2\right)\right] \quad (2.2)$$

Let us prove the proposition for $a_1 - a_L > 0$, $b + b_R < 0$ ($\tilde{A}$ — a positive number, $\tilde{B}$ — a negative number). The proof of the proposition for other relations between numbers $\tilde{A}, \tilde{B}$ is carried out similarly. In a case $a_1 - a_L > 0$, $b + b_R < 0$ for (2.2) we'll obtain

$$D_\alpha = \left[D_\alpha^1, D_\alpha^2\right] =$$
$$= \left\{\left[a_2 + R_1^{-1}(\alpha)a_{R_1}\right]\left[b_1 - L_2^{-1}(\alpha)b_{L_2}\right], \left[a_1 - L_1^{-1}(\alpha)a_{L_1}\right]\left[b_2 + R_2^{-1}(\alpha)b_{R_2}\right]\right\}.$$

Functions L_1, L_2, R_1, R_2 are monotonically decreasing, therefore at $a_1 - a_L > 0$, $b + b_R < 0$ and $\alpha_2 > \alpha_1$

$$0 < a_1 - L_1^{-1}(\alpha_1)a_{L_1} < a_1 - L_1^{-1}(\alpha_2)a_{L_1};$$

$$0 < a_2 + R_1^{-1}(\alpha_2)a_{R_1} < a_2 + R_1^{-1}(\alpha_1)a_{R_1};$$

$$b_1 - L_2^{-1}(\alpha_1)b_{L_2} < b_1 - L_2^{-1}(\alpha_2)b_{L_2} < 0;$$

$$b_2 + R_2^{-1}(\alpha_2)b_{R_2} < b_2 + R_2^{-1}(\alpha_1)b_{R_2} < 0.$$

Thus, with $\alpha_2 > \alpha_1$ $D_{\alpha 2}^1 > D_{\alpha 1}^1$, $D_{\alpha 2}^2 > D_{\alpha 1}^2$, and hence, the left boundary of membership function $\tilde{D}$ monotonically increases, and the right monotonically decreases, that ensures a membership of $\tilde{D}$ to $\wedge$. The proposition 2.1 is proved.

Similarly, it is possible to prove that product of $\wedge$-unimodal numbers is $\wedge$-unimodal number, and product of $\wedge$-tolerance and $\wedge$-unimodal numbers is $\wedge$-tolerance number.

Further paragraphs of the Chapter are devoted to methods of formalization of expert evaluations of qualitative characteristics and description of values of quantitative characteristics in linguistic terms based on COSS. The constructed population of fuzzy numbers of $\wedge$ is offered to be used for formalization of linguistic values of these characteristics or for building of membership functions of COSS terms.

2.2 Formalization of Expert Evaluations of Qualitative Characteristics in a Verbal Scale

As the input information the a posteriori information resulted from expert evaluations of a qualitative characteristic X for a population of objects [121—122] is considered.

It is assumed that the evaluation was carried out within the limits of a verbal scale with levels X_l, $l = \overline{1, m}$; $m \geq 2$ ordered on increase of characteristic appearance intensity degree. Levels of a used verbal scale unambiguously specify term-set $T(X) = \{X_1, X_2, ..., X_m\}$. $U = [0,1]$ is selected as universal COSS set. The point $x = 0$ corresponds to a total absence of characteristic X appearance, and consequently it is considered as a typical point of the term X_1. The point $x = 1$ corresponds to total presence of characteristic X appearance, and consequently it is considered as a typical point of term X_m.

Let us denote relative frequencies of objects' occurrence for which characteristic X intensity is estimated by levels x_l by a_l, accordingly

$$\sum_{l=1}^{m} a_l = 1.$$

Let us assume that fuzzy numbers corresponding to terms X_l with membership functions $\mu_l(x)$ belong to population Λ and satisfy a side condition (1*):

If $L(x)$, $R(x)$ are nonlinear functions, they have a central symmetry versus point of inflexion.

Building of membership functions of COSS term-set will be carried out so that squares of the figures limited to graphs of functions $\mu_l(x)$ and an axis of abscissas be equal to a_l. It is obvious that the there is an infinite number of membership functions meeting such requirements, therefore it is necessary to limit building to logical requirements for fuzziness areas between the adjacent terms (or to parameters of fuzziness of the fuzzy numbers corresponding to terms). On the one hand, there is a desire to make this area as small as possible, then the degree of fuzziness (1.7) constructed as a COSS model will be less accordingly. On the other hand, it is necessary to rest upon substantial sense of fuzziness area, so we propose to calculate a potency (length) of this area for extreme terms as $\min(a_1, a_2)$ or $\min(a_{m-1}, a_m)$, accordingly, and for mean terms to calculate the same base the relations between numbers a_{l-1}, a_l, a_{l+1}; $l = \overline{3; m-2}$. Graphs of the constructed membership functions will be in a form of curvilinear trapezoids with midlines equal to a_l, $l = \overline{1, m}$.

Let us construct membership function of term X_m:

1. If $a_m \leq a_{m-1}$, then

$$\mu_m(x) = \left\{ L \begin{array}{ll} 0, & 0 \le x \le \left(1 - \dfrac{3a_m}{2}\right); \\[3mm] \dfrac{1 - \dfrac{a_m}{2} - x}{a_m}, & \left(1 - \dfrac{3a_m}{2}\right) < x \le \left(1 - \dfrac{a_m}{2}\right); \\[3mm] 1, & \left(1 - \dfrac{a_m}{2}\right) < x \le 1. \end{array} \right.$$

Fig. 2.1 shows membership functions of term X_m for a special case $L(x) = 1 - x$; $0 \le x \le 1$ (membership functions of T-numbers). The special attention is given to this special case because of large number of application of T-numbers and normal triangular numbers in applied researches.

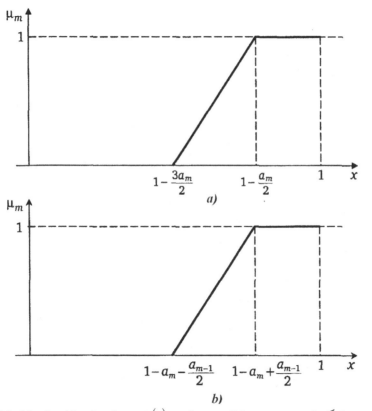

Fig. 2.1 Membership function $\mu_m(x)$ under conditions: a — $a_m \le a_{m-1}$, b — $a_m > a_{m-1}$

2. If $a_m > a_{m-1}$, then

$$\mu_m(x) = \begin{cases} 0, & 0 \le x \le \left(1 - a_m - \dfrac{a_{m-1}}{2}\right); \\ L\left(\dfrac{1 - a_m + \dfrac{a_{m-1}}{2} - x}{a_{m-1}}\right), & \left(1 - a_m - \dfrac{a_{m-1}}{2}\right) < x \le \left(1 - a_m + \dfrac{a_{m-1}}{2}\right); \\ 1, & \left(1 - a_m + \dfrac{a_{m-1}}{2}\right) < x \le 1. \end{cases}$$

Let us construct membership function of term X_{m-1}:

1. If $a_{m-1} \ge \max(a_m, a_{m-2})$, then

$$\mu_{m-1}(x) = \begin{cases} 0, & 0 \le x \le \left(\displaystyle\sum_{l=1}^{m-3} a_l + \dfrac{a_{m-2}}{2}\right); \\ R\left(\dfrac{\displaystyle\sum_{l=1}^{m-2} a_l + \dfrac{a_{m-2}}{2} - x}{a_{m-2}}\right), & \left(\displaystyle\sum_{l=1}^{m-3} a_l + \dfrac{a_{m-2}}{2}\right) < x \le \left(\displaystyle\sum_{l=1}^{m-2} a_l + \dfrac{a_{m-2}}{2}\right); \\ 1, & \left(\displaystyle\sum_{l=1}^{m-2} a_l + \dfrac{a_{m-2}}{2}\right) < x \le \left(1 - \dfrac{3a_m}{2}\right); \\ L\left(\dfrac{x - 1 + \dfrac{3a_m}{2}}{a_m}\right), & \left(1 - \dfrac{3a_m}{2}\right) < x \le \left(1 - \dfrac{a_m}{2}\right); \\ 0, & \left(1 - \dfrac{a_m}{2}\right) < x \le 1. \end{cases}$$

Fig. 2.2 shows membership functions of a term X_{m-1} for the special case $L(x) = 1 - x$; $R(x) = 1 - x$; $0 \le x \le 1$.

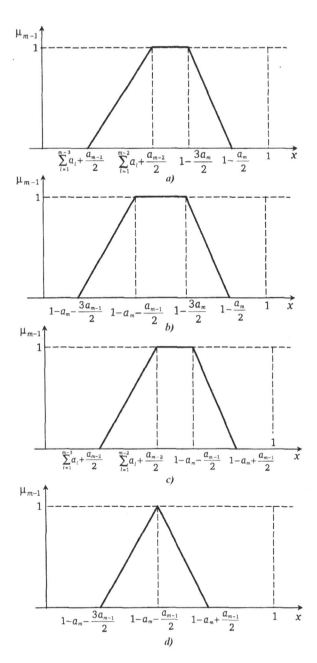

Fig. 2.2 Membership function μ_{m-1} under conditions: a — $a_{m-1} \geq \max(a_m, a_{m-2})$, b — $a_m < a_{m-1} < a_{m-2}$, c — $a_{m-2} < a_{m-1} < a_m$, d — $a_{m-1} \leq \min(a_m, a_{m-2})$

2. If $a_m < a_{m-1} < a_{m-2}$, then

$$\mu_{m-1}(x) = \begin{cases} 0, & 0 \le x \le \left(1 - a_m - \dfrac{3a_{m-1}}{2}\right); \\[3mm] R\left(\dfrac{1 - a_m - \dfrac{a_{m-1}}{2} - x}{a_{m-1}}\right), & \left(1 - a_m - \dfrac{3a_{m-1}}{2}\right) < x \le \left(1 - a_m - \dfrac{a_{m-1}}{2}\right); \\[3mm] 1, & \left(1 - a_m - \dfrac{a_{m-1}}{2}\right) < x \le \left(1 - \dfrac{3a_m}{2}\right); \\[3mm] L\left(\dfrac{x - 1 + \dfrac{3a_m}{2}}{a_m}\right), & \left(1 - \dfrac{3a_m}{2}\right) < x \le \left(1 - \dfrac{a_m}{2}\right); \\[3mm] 0, & \left(1 - \dfrac{a_m}{2}\right) < x \le 1. \end{cases}$$

3. If $a_{m-2} < a_{m-1} < a_m$, then

$$\mu_{m-1}(x) = \begin{cases} 0, & 0 \le x \le \left(\sum\limits_{l=1}^{m-3} a_l + \dfrac{a_{m-2}}{2}\right); \\[3mm] R\left(\dfrac{\sum\limits_{l=1}^{m-2} a_l + \dfrac{a_{m-2}}{2} - x}{a_{m-2}}\right), & \left(\sum\limits_{l=1}^{m-3} a_l + \dfrac{a_{m-2}}{2}\right) < x \le \left(\sum\limits_{l=1}^{m-2} a_l + \dfrac{a_{m-2}}{2}\right); \\[3mm] 1, & \left(\sum\limits_{l=1}^{m-2} a_l + \dfrac{a_{m-2}}{2}\right) < x \le \left(1 - a_m - \dfrac{a_{m-1}}{2}\right); \\[3mm] L\left(\dfrac{x - 1 + a_m + \dfrac{a_{m-1}}{2}}{a_{m-1}}\right), & \left(1 - a_m - \dfrac{a_{m-1}}{2}\right) < x \le \left(1 - a_m + \dfrac{a_{m-1}}{2}\right); \\[3mm] 0, & \left(1 - a_m + \dfrac{a_{m-1}}{2}\right) < x \le 1. \end{cases}$$

4. If $a_{m-1} \leq \min(a_m, a_{m-2})$, then

$$\mu_{m-1}(x) = \begin{cases} 0, & 0 \leq x \leq \left(1 - a_m - \dfrac{3a_{m-1}}{2}\right); \\[4mm] R\left(\dfrac{1 - a_m - \dfrac{a_{m-1}}{2} - x}{a_{m-1}}\right), & \left(1 - a_m - \dfrac{3a_{m-1}}{2}\right) < x \leq \left(1 - a_m - \dfrac{a_{m-1}}{2}\right); \\[4mm] L\left(\dfrac{x - 1 + a_m + \dfrac{a_{m-1}}{2}}{a_{m-1}}\right), & \left(1 - a_m - \dfrac{a_{m-1}}{2}\right) < x \leq \left(1 - a_m + \dfrac{a_{m-1}}{2}\right); \\[4mm] 0, & \left(1 - a_m + \dfrac{a_{m-1}}{2}\right) < x \leq 1; \end{cases}$$

Similarly to $\mu_{m-1}(x)$ membership functions $\mu_l(x)$; $l = \overline{2, m-2}$ are constructed.

Let us construct membership function for term X_1 with even number of terms.

1. If $a_1 \leq a_2$, then

$$\mu_1(x) = \begin{cases} 1, & 0 \leq x \leq \dfrac{a_1}{2}; \\[4mm] L\left(\dfrac{x - \dfrac{a_1}{2}}{a_1}\right), & \dfrac{a_1}{2} < x \leq \dfrac{3a_1}{2}; \\[4mm] 0, & \dfrac{3a_1}{2} < x \leq 1. \end{cases}$$

2. If $a_1 > a_2$, then

$$\mu_1(x) = \begin{cases} 1, & 0 \leq x \leq \left(a_1 - \dfrac{a_2}{2}\right); \\[4mm] L\left(\dfrac{x - a_1 + \dfrac{a_2}{2}}{a_2}\right), & \left(a_1 - \dfrac{a_2}{2}\right) < x \leq \left(a_1 + \dfrac{a_2}{2}\right); \\[4mm] 0, & \left(a_1 + \dfrac{a_2}{2}\right) < x \leq 1. \end{cases}$$

With odd number of terms we obtain:

1. If $a_1 \leq a_2$, then

$$\mu_1(x) = \begin{cases} 1, & 0 \leq x \leq \dfrac{a_1}{2}; \\[3ex] R\left(\dfrac{x - \dfrac{a_1}{2}}{a_1}\right), & \dfrac{a_1}{2} < x \leq \dfrac{3a_1}{2}; \\[3ex] 0, & \dfrac{3a_1}{2} < x \leq 1. \end{cases}$$

2. If $a_1 > a_2$, then

$$\mu_1(x) = \begin{cases} 1, & 0 \leq x \leq \left(a_1 - \dfrac{a_2}{2}\right); \\[3ex] R\left(\dfrac{x - a_1 + \dfrac{a_2}{2}}{a_2}\right), & \left(a_1 - \dfrac{a_2}{2}\right) < x \leq \left(a_1 + \dfrac{a_2}{2}\right); \\[3ex] 0, & \left(a_1 + \dfrac{a_2}{2}\right) < x \leq 1. \end{cases}$$

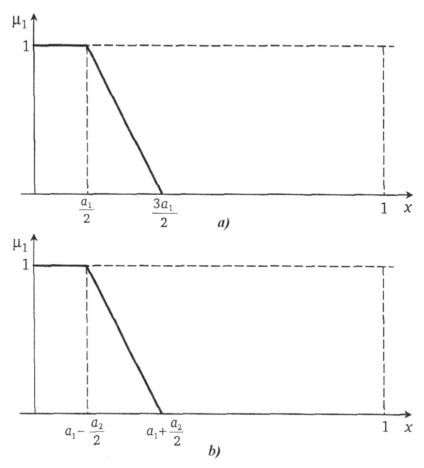

Fig. 2.3 Membership function $\mu_1(x)$ under conditions: **a** — $a_1 \le a_2$; **b** — $a_1 > a_2$

Fig. 2.3 shows membership functions for special case $L(x)=1-x$; $0 \le x \le 1$.

Let us construct membership functions of COSS terms with linear boundaries, i.e. we assume that $L(x) = R(x) = 1-x$. Graphs of these membership functions for various relations between $a_l \left(l = \overline{1, m}\right)$ are shown in Fig. 2.1 – 2.3.

Let us denote $min\left(a_l, a_{l+1}\right)$, $l = \overline{1, m-1}$ by b_l.

Then

$$\mu_1(x)=\begin{cases} 1, & 0\le x\le\left(a_1-\dfrac{b_1}{2}\right); \\[3mm] 1-\dfrac{x-\left(a_1-\dfrac{b_1}{2}\right)}{b_1}, & \left(a_1-\dfrac{b_1}{2}\right)<x\le\left(a_1+\dfrac{b_1}{2}\right); \\[3mm] 0, & \left(a_1+\dfrac{b_1}{2}\right)<x\le1; \end{cases}$$

$$\mu_l(x)=\begin{cases} 0, & 0\le x\le\left(\displaystyle\sum_{i=1}^{l-1}a_i-\dfrac{b_{l-1}}{2}\right); \\[4mm] 1+\dfrac{x-\left(\displaystyle\sum_{i=1}^{l-1}a_i+\dfrac{b_{l-1}}{2}\right)}{b_{l-1}}, & \left(\displaystyle\sum_{i=1}^{l-1}a_i-\dfrac{b_{l-1}}{2}\right)<x\le\left(\displaystyle\sum_{i=1}^{l-1}a_i+\dfrac{b_{l-1}}{2}\right); \\[4mm] 1, & \left(\displaystyle\sum_{i=1}^{l-1}a_i+\dfrac{b_{l-1}}{2}\right)<x\le\left(\displaystyle\sum_{i=1}^{l}a_i-\dfrac{b_l}{2}\right); \\[4mm] 1-\dfrac{x-\left(\displaystyle\sum_{i=1}^{l}a_i-\dfrac{b_l}{2}\right)}{b_l}, & \left(\displaystyle\sum_{i=1}^{l}a_i-\dfrac{b_l}{2}\right)<x\le\left(\displaystyle\sum_{i=1}^{l}a_i+\dfrac{b_l}{2}\right); \\[4mm] 0, & \left(\displaystyle\sum_{i=1}^{l}a_i+\dfrac{b_l}{2}\right)<x\le1; \end{cases}$$

$$\mu_{m-1}(x)=\begin{cases} 0, & 0\le x\le\left(\displaystyle\sum_{i=1}^{m-2}a_i-\dfrac{b_{m-2}}{2}\right); \\[4mm] 1+\dfrac{x-\left(\displaystyle\sum_{i=1}^{m-2}a_i+\dfrac{b_{l-1}}{2}\right)}{b_{m-2}}, & \left(\displaystyle\sum_{i=1}^{m-2}a_i-\dfrac{b_{m-2}}{2}\right)<x\le\left(\displaystyle\sum_{i=1}^{m-2}a_i+\dfrac{b_{m-2}}{2}\right); \\[4mm] 1, & \left(\displaystyle\sum_{i=1}^{m-2}a_i+\dfrac{b_{m-2}}{2}\right)<x\le\left(1-a_m-\dfrac{b_{m-1}}{2}\right); \\[4mm] 1-\dfrac{x-\left(1-a_m-\dfrac{b_{m-1}}{2}\right)}{b_{m-1}}, & \left(1-a_m-\dfrac{b_{m-1}}{2}\right)<x\le\left(1-a_m+\dfrac{b_{m-1}}{2}\right); \\[4mm] 0, & \left(1-a_m+\dfrac{b_{m-1}}{2}\right)<x\le1; \end{cases}$$

$$\mu_m(x) = \begin{cases} 0, & 0 \le x \le \left(1 - a_m - \dfrac{b_{m-1}}{2}\right); \\[3mm] 1 + \dfrac{x - \left(1 - a_m + \dfrac{b_{m-1}}{2}\right)}{b_{m-1}}, & \left(1 - a_m - \dfrac{b_{m-1}}{2}\right) < x \le \left(1 - a_m + \dfrac{b_{m-1}}{2}\right); \\[3mm] 1, & \left(1 - a_m + \dfrac{b_{m-1}}{2}\right) < x \le 1. \end{cases}$$

Logic of requirements to fuzziness area imposed to construct COSS term-set membership functions can be explained with the following simple example: when a mean evaluation of knowledge (certificate, diploma etc.) is identified with one of the accepted marks «2», «3», «4», «5», the mean evaluation which falls within an interval (4; 4.5), for example, is identified with a mark «4», and the mean evaluation which falls within an interval [4.5; 5] is identified with a mark «5». If an expert has his/her own judgments concerning fuzziness areas between the adjacent terms, then the building method is corrected and put into practice. If an expert agrees with the interpretation of authors stated herein, or information is processed by a person who does not possess expert experience, the information is proposed for use without any modifications.

The described building of COSS membership functions is offered to be applied not only in the conditions of availability of a posteriori information submitted to processing. The expert can build COSS without such information at the moment using the information which he/she enjoyed earlier owing to considerable experience gained.

To estimate qualitative characteristics, verbal-numerical or numerical ordinal scales can be used. The model-building techniques of COSS membership functions with the use of these scales is invariable.

Example 2.1. Model-building of COSS "quality of production". A firm manufactures production of the superior, first, second and third quality degrees. Over a certain period, 523 production units of the superior quality, 1084 production units of the first quality, 857 production units of the second quality and 379 production units of the third quality are manufactured. Let us construct COSS "quality of production" with terms $X_1 = \{$the third quality$\}$, $X_2 = \{$the second quality$\}$, $X_3 = \{$the first quality$\}$, $X_4 := \{$the superior quality$\}$. Let us denote with $\mu_1(x)$ a membership function of X_1, with $\mu_2(x)$ - a membership function of X_2, with $\mu_3(x)$ - a membership function of X_3, and with $\mu_4(x)$ - a membership function for X_4. Let us obtain relative frequencies for production units of the first - third and superior qualities: $a_1 = 0{,}133$, $a_2 = 0{,}301$, $a_3 = 0{,}382$, $a_4 = 0{,}184$, accordingly. Then

$$\mu_1(x) = \begin{cases} 1, & 0 \le x \le 0{,}0665; \\ 1 - \dfrac{x - 0{,}0665}{0{,}133}, & 0{,}0665 < x \le 0{,}1995; \\ 0, & 0{,}1995 < x \le 1; \end{cases}$$

$$\mu_2(x) = \begin{cases} 0, & x \le 0{,}0665; \\ 1 + \dfrac{x - 0{,}1995}{0{,}133}, & 0{,}0665 < x \le 0{,}1995; \\ 1, & 0{,}1995 < x \le 0{,}2835; \\ 1 - \dfrac{x - 0{,}2835}{0{,}301}, & 0{,}2835 < x \le 0{,}5845; \\ 0, & 0{,}5845 < x \le 1; \end{cases}$$

$$\mu_3(x) = \begin{cases} 0, & 0 \le x \le 0{,}2835; \\ 1 + \dfrac{x - 0{,}5845}{0{,}301}, & 0{,}2835 < x \le 0{,}5845; \\ 1, & 0{,}5845 < x \le 0{,}724; \\ 1 - \dfrac{x - 0{,}724}{0{,}184}, & 0{,}724 < x \le 0{,}908; \\ 0, & 0{,}908 < x \le 1; \end{cases}$$

$$\mu_4(x) = \begin{cases} 1, & 0 \le x \le 0{,}724; \\ 1 + \dfrac{x - 0{,}908}{0{,}184}, & 0{,}724 < x \le 0{,}908; \\ 0, & 0{,}908 < x \le 1. \end{cases}$$

Graphs of the obtained membership functions are shown in Fig. 2.4. a.

Example 2.2. Model-building of COSS "knowledge of students".
There are data of progress of 100 students of a certain specialty over a certain period of time

Mark	Number of students
E ("unsatisfactory")	10
C ("satisfactory")	40
B ("good")	30
A ("excellent")	20

Let us construct COSS "knowledge of students" with terms "E","C", "B", "A" and membership functions $\mu_{\tilde{2}}(x) - \mu_{\tilde{5}}(x)$, accordingly. $T(x) = \{2, 3, 4, 5\}$. Based on values of relative frequencies, we obtain:

$$\mu_{\tilde{2}}(x) = \begin{cases} 1, & 0 \le x \le 0{,}05; \\ 1 - \dfrac{x - 0{,}05}{0{,}1}, & 0{,}05 < x \le 0{,}15; \\ 0, & 0{,}15 < x \le 1, \end{cases}$$

$$\mu_{\tilde{3}}(x) = \begin{cases} 0, & x \le 0,05; \\ 1 + \dfrac{x - 0,15}{0,1}, & 0,05 < x \le 0,15; \\ 1, & 0,15 < x \le 0,35; \\ 1 - \dfrac{x - 0,35}{0,3}, & 0,35 < x \le 0,65; \\ 0, & 0,65 < x \le 1; \end{cases}$$

$$\mu_{\tilde{4}}(x) = \begin{cases} 0, & 0 \le x \le 0,35; \\ 1 + \dfrac{x - 0,65}{0,3}, & 0,35 < x \le 0,65; \\ 1, & 0,65 < x \le 0,7; \\ 1 - \dfrac{x - 0,7}{0,2}, & 0,7 < x \le 0,9; \\ 0, & 0,9 < x \le 1; \end{cases}$$

$$\mu_{\tilde{5}}(x) = \begin{cases} 1, & 0 \le x \le 0,7; \\ 1 + \dfrac{x - 0,9}{0,2}, & 0,7 < x \le 0,9; \\ 0, & 0,9 < x \le 1. \end{cases}$$

Graphs of the obtained membership functions are shown in Fig. 2.4.b.

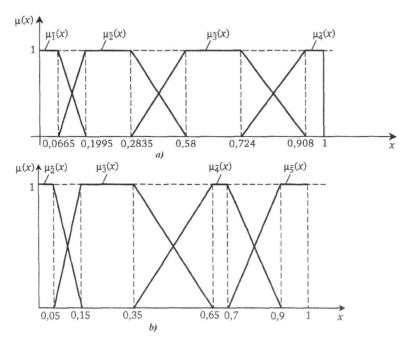

Fig. 2.4 Membership functions of COSS terms: a— "quality of production"; b — "knowledge of students"

2.3 Formalization of Expert Evaluations of Qualitative Characteristics in a Mark Scale

Let us assume that appearances (demonstrations) of qualitative characteristics are estimated by experts using a mark scale.

Let us construct COSS on the basis of a posteriori information resulted from an evaluation of appearances of qualitative characteristic X for a population of objects $Y_1, Y_2, ..., Y_N$. The minimum quantity of points which allows evaluation of characteristic appearance equals zero, and maximum quantity of points is M. Besides it is assumed that the verbal scale is developed for the evaluation purposes, with levels X_l, $l = \overline{1, m}$ arranged in ascending intensity order.

Let $y_j \in [0,1]$, $j = \overline{1, N}$ represent relative evaluations of appearance of characteristic X, which were obtained by dividing mark scale-based evaluations by the maximum estimate M. For example, in terms of educational process evaluation they can be obtained as a result of testing within the scope of a certain school subject, and those evaluation values are equal to the ratio of number of properly performed tasks to the total quantity of all tasks performed.

Let us construct COSS X with terms X_l (according to levels of a verbal scale) and membership functions $\mu_l(x)$ of T-numbers or normal triangular numbers.

The additional expert information consisting of results of preliminary allocation of outcomes obtained to one of levels of a verbal scale and comparison of outcomes between each other is required for model-building process. In this case, the standard approach of paired comparisons of outcomes of qualitative characteristics availability, which is applied to construct membership functions of fuzzy sets [49], is used.

Let us start constructing with term X_m which corresponds to maximum characteristic X appearance intensity degree. Let the point $x = 1$, i.e. $\mu_m(1) = 1$ be considered as the typical one for the membership function of this term.

Let $Y_1, ..., Y_j$; $j \geq 3$ are the objects allocated by the expert to level X_m. Let us arrange them in decreasing order of evaluations y_i; $i = \overline{1, j}$. We obtain a conditional ordered series

$$Y_{(1)}, Y_{(2)}, ..., Y_{(j)},$$

to which a numerical ordered series corresponds

$$y_{(1)}, y_{(2)}, ..., y_{(j)}.$$

Let us perform paired comparisons of objects of that conditional ordered series using Saati scale [123]. Let $a_{ik} = \overline{1, 9}$, $k = \overline{1, j}$ be Saati scale evaluations of characteristic X superiority availability at an object $X_{(i)}$ versus an object $Y_{(k)}$. If objects $Y_{(i)}$ and $Y_{(k)}$ have approximately equal characteristic X appearances, then

$a_{ik} = 1$; if the object $Y_{(i)}$ appearances slightly exceeds those of the object $Y_{(k)}$, then $a_{ik} = 3$. If the characteristic X appearances of the object $Y_{(i)}$ exceed those of the object $Y_{(k)}$ to the extent "more", "noticeably more" or "much more", then $a_{ik} = 5; 7; 9$. Evaluations $a_{ik} = 2; 4; 6; 8$ are intermediate.

Let us assume $a_{ki} = 1/a_{ik}$, compose a matrix of paired comparisons

$$A_m = \begin{pmatrix} 1 & a_{12} & a_{13} & \cdots & a_{1j} \\ \dfrac{1}{a_{12}} & 1 & a_{23} & \cdots & a_{2j} \\ \dfrac{1}{a_{13}} & \dfrac{1}{a_{23}} & 1 & \cdots & a_{3j} \\ \vdots & \vdots & \vdots & \cdots & \vdots \\ \dfrac{1}{a_{1j}} & \dfrac{1}{a_{2j}} & \dfrac{1}{a_{3j}} & \cdots & 1 \end{pmatrix}$$

and determine its eigenvalues. For this purpose let us equate a determinant to zero.

$$\begin{vmatrix} 1-\lambda & a_{12} & a_{13} & \cdots & a_{1j} \\ \dfrac{1}{a_{12}} & 1-\lambda & a_{23} & \cdots & a_{2j} \\ \dfrac{1}{a_{13}} & \dfrac{1}{a_{23}} & 1-\lambda & \cdots & a_{3j} \\ \vdots & \vdots & \vdots & \cdots & \vdots \\ \dfrac{1}{a_{1j}} & \dfrac{1}{a_{2j}} & \dfrac{1}{a_{3j}} & \cdots & 1-\lambda \end{vmatrix} = 0.$$

Let us choose maximum eigenvalue λ_{max} and determine a corresponding eigenvector $\omega_m = (\omega_{m,1}, \ldots, \omega_{m,j})$. For this purpose we'll solve a set of equations written in the matrix form

$$\begin{pmatrix} 1-\lambda_{max} & a_{12} & a_{13} & \cdots & a_{1j} \\ \dfrac{1}{a_{12}} & 1-\lambda_{max} & a_{23} & \cdots & a_{2j} \\ \dfrac{1}{a_{13}} & \dfrac{1}{a_{23}} & 1-\lambda_{max} & \cdots & a_{3j} \\ \vdots & \vdots & \vdots & \cdots & \vdots \\ \dfrac{1}{a_{1j}} & \dfrac{1}{a_{2j}} & \dfrac{1}{a_{3j}} & \cdots & 1-\lambda_{max} \end{pmatrix} \begin{pmatrix} \omega_{m,1} \\ \omega_{m,2} \\ \omega_{m,3} \\ \vdots \\ \omega_{m,j} \end{pmatrix} = 0.$$

(2.3)

It is known [15], that the system (2.3) has a solution and

$$\omega_{m,i} > 0.$$

If eigenvector co-ordinates do not belong to a segment [0. 1], we'll determine

$$\tilde{\omega}_m = \sum_{i=1}^{j} \omega_{m,i}$$

and we obtain

$$\tilde{\omega}_{m,j} = \frac{\omega_{m,i}}{\tilde{\omega}_m}.$$

We consider $\tilde{\omega}_{m,i}$ as grade of membership of the object $Y_{(i)}$ to term X_m. As for each object $Y_{(i)}$ evaluations $y(i) \in [0,1]$ are determined, let us consider that they belong to term X_m with membership degrees $\tilde{\omega}_{m,i}$, accordingly.

In all further buildings the assumption is made that co-ordinates of the eigenvector corresponding to a maximum eigenvalue belong to the segment [0, 1], otherwise normalization is used.

In order to obtain membership function $\mu_m(x)$ of the term X_m the left boundary of which looks like $y = a_m x + b_m$, and right, in turn, is $x = 1$, let us use a method of least squares

$$F_m = \sum_{i=1}^{j} \left(a_m y_{(i)} + b_m - \omega_{m,i} \right)^2 \to \min.$$

From the system of normal equations

$$
\begin{cases}
\dfrac{\partial F_m}{\partial a_m} = 0 \\
\dfrac{\partial F_m}{\partial b_m} = 0
\end{cases}
\Leftrightarrow
\begin{cases}
a_m \sum_{i=1}^{j} y_{(i)}^2 + b_m \sum_{i=1}^{j} y_{(i)} = \sum_{i=1}^{j} y_{(i)} \omega_{m,i}; \\
a_m \sum_{i=1}^{j} y_{(i)} + j b_m = \sum_{i=1}^{j} \omega_{m,i}.
\end{cases}
$$

we obtain unknown coefficients:

$$b_m = \frac{j \sum_{i=1}^{j} y_{(i)} \omega_{m,i} - \left(\sum_{i=1}^{j} y_{(i)} \right)\left(\sum_{i=1}^{j} \omega_{m,i} \right)}{j \sum_{i=1}^{j} y_{(i)}^2 - \left(\sum_{i=1}^{j} y_{(i)} \right)^2};$$

$$a_m = \frac{1}{j} \sum_{i=1}^{j} \omega_{m,i} - \frac{1}{j} \sum_{i=1}^{j} y_{(i)} \frac{j \sum_{i=1}^{j} y_{(i)} \omega_{m,i} - \left(\sum_{i=1}^{j} y_{(i)} \right)\left(\sum_{i=1}^{j} \omega_{m,i} \right)}{j \sum_{i=1}^{j} y_{(i)}^2 - \left(\sum_{i=1}^{j} y_{(i)} \right)^2}.$$

If the condition $y(1) = a_m + b_m > 1$ is satisfied, membership function of term X_m will be membership function of T -number:

$$\mu_m(x) = \begin{cases} 0, & 0 \leq x \leq -\dfrac{b_m}{a_m}; \\ a_m x + b_m, & -\dfrac{b_m}{a_m} < x \leq \dfrac{1 - b_m}{a_m}; \\ 1, & \dfrac{1 - b_m}{a_m} < x \leq 1. \end{cases}$$

If $y(1) = a_m + b_m \leq 1$, the left boundary of membership function is obtained as $y = a_m x + 1 - a_m$, and the unknown coefficient a_m is obtained from an optimization problem solved

$$F_m = \sum_{i=1}^{j} \left(a_m y_{(i)} + 1 - a_m - \omega_{m,i} \right)^2 \rightarrow \min;$$

$$\frac{\partial F_m}{\partial a_m} = 0 \Leftrightarrow a_m = \frac{\sum_{i=1}^{j} \left(y_{(i)} - 1 \right) \left(\omega_{m,i} - 1 \right)}{\sum_{i=1}^{j} \left(y_{(i)} - 1 \right)^2}.$$

In this case, membership function of term X_m will be membership function of a normal triangular number:

$$\mu_m(x) = \begin{cases} 0, & 0 \leq x \leq \dfrac{a_m - 1}{a_m}; \\ a_m x + 1 - a_m, & \dfrac{a_m - 1}{a_m} < x \leq 1. \end{cases}$$

Definition of the left boundary of membership function of term X_m yields an unambiguous definition of the right boundary of membership function of term X_{m-1}, i.e.

with $-\dfrac{b_m}{a_m} < x \leq \dfrac{1 - b_m}{a_m}$

$$\mu_{m-1}(x) = 1 - a_m x - b_m.$$

Let us consider the objects allocated by the expert to level X_{m-1}. Let $Y_{j+1}, \ldots, Y_{j+v}$ $(v \geq 3)$ be the objects allocated by the expert to level X_{m-1}, then let us arrange them in decreasing order of evaluations y_i $(i = \overline{j+1; j+v})$. We obtain a conditional ordered series

$$Y_{(j+1)}, Y_{(j+2)}, ..., Y_{(j+v)},$$

to which, as we see above, the numerical ordered series corresponds

$$y_{(j+1)}, y_{(j+2)}, ..., y_{(j+v)},$$

Let us perform paired comparisons of objects of this series using Saati scale, determine a matrix of paired comparisons and its eigenvector $\omega_{m-1} = (\omega_{m-1,1}, ..., \omega_{m-1,v})$ corresponding to a maximum eigenvalue. Let us consider that evaluations $y_i \in [0,1]$, $i = \overline{j+1; j+v}$ belong to term X_{m-1} with membership degrees ω_{m-1}, $i = \overline{1,v}$, accordingly. To obtain membership function $\mu_{m-1}(x)$, with the left boundary in the form $y = a_{m-1}x + b_{m-1}$, let us use a method of least squares:

$$F_{m-1} = \sum_{i=1}^{v} \left(a_{m-1} y_{(j+1)} + b_{m-1} - \omega_{m-1,i} \right)^2 \rightarrow \min.$$

From system of normal equations

$$\begin{cases} \dfrac{\partial F_{m-1}}{\partial a_{m-1}} = 0; \\ \dfrac{\partial F_{m-1}}{\partial b_{m-1}} = 0 \end{cases}$$

We obtain unknown coefficients

$$b_{m-1} = \frac{v \sum\limits_{i=1}^{v} y_{(j+i)} \omega_{m-1,i} - \left(\sum\limits_{i=1}^{v} y_{(j+i)} \right) \left(\sum\limits_{i=1}^{v} \omega_{m-1,i} \right)}{v \sum\limits_{i=1}^{v} y_{(j+i)}^2 - \left(\sum\limits_{i=1}^{v} y_{(j+i)} \right)^2};$$

$$a_{m-1} = \frac{1}{v} \sum_{i=1}^{v} \omega_{m-1,i} - \frac{1}{v} \sum_{i=1}^{v} y_{(j+i)} \frac{v \sum\limits_{i=1}^{v} y_{(j+i)} \omega_{m-1,i} - \left(\sum\limits_{i=1}^{v} y_{(j+i)} \right) \left(\sum\limits_{i=1}^{v} \omega_{m-1,i} \right)}{v \sum\limits_{i=1}^{v} y_{(j+i)}^2 - \left(\sum\limits_{i=1}^{v} y_{(j+i)} \right)^2}.$$

If the condition

$$y\left(-\frac{b_m}{a_m} \right) = -a_{m-1} \frac{b_m}{a_m} + b_{m-1} > 1,$$

is satisfied, we obtain membership function of T-number:

$$\mu_{m-1}(x) = \begin{cases} 0, & 0 \le x \le -\dfrac{b_{m-1}}{a_{m-1}}; \\[2mm] a_{m-1}x + b_{m-1}, & -\dfrac{b_{m-1}}{a_{m-1}} < x \le \dfrac{1-b_{m-1}}{a_{m-1}}; \\[2mm] 1, & \dfrac{1-b_{m-1}}{a_{m-1}} < x \le -\dfrac{b_m}{a_m}; \\[2mm] 1 - a_m x - b_m, & -\dfrac{b_m}{a_m} < x \le \dfrac{1-b_m}{a_m}; \\[2mm] 0, & \dfrac{1-b_m}{a_m} < x \le 1. \end{cases}$$

if this condition is not satisfied, it is supposed that

$$-a_{m-1}\frac{b_m}{a_m} + b_{m-1} = 1.$$

The unknown coefficient a_{m-1} is obtained from the condition

$$F_{m-1} = \sum_{i=1}^{v}\left(a_{m-1}y_{(j+1)} + b_{m-1} - \omega_{m-1,i}\right)^2 \to \min.$$

In this case we obtain membership function of a normal triangular number

$$\mu_{m-1}(x) = \begin{cases} 0, & 0 \le x \le \left(-\dfrac{a_m + a_{m-1}b_m}{a_m a_{m-1}}\right); \\[2mm] a_{m-1}x + 1 + a_{m-1}\dfrac{b_m}{a_m}, & \left(-\dfrac{a_m + a_{m-1}b_m}{a_m a_{m-1}}\right) < x \le -\dfrac{b_m}{a_m}; \\[2mm] 1 - a_m x - b_m, & -\dfrac{b_m}{a_m} < x \le \dfrac{1-b_m}{a_m}; \\[2mm] 0, & \dfrac{1-b_m}{a_m} < x \le 1. \end{cases}$$

Definition of the left boundary of membership function of term X_{m-1} yields an unambiguous definition of its right boundary $\mu_{m-2}(x)$ of term X_{m-2}, i.e. at $-\dfrac{b_{m-1}}{a_{m-1}} < x \le \dfrac{1-b_{m-1}}{a_{m-1}}$

$$\mu_{m-2}(x) = 1 - a_{m-1}x - b_{m-1}.$$

The membership functions $\mu_l(x)$; $l = \overline{3, m-2}$ is constructed as described above.

Let us see in detail building process of membership functions $\mu_1(x)$ and $\mu_2(x)$. The left boundary of membership function $\mu_3(x)$ unambiguously defines the right boundary of membership function $\mu_2(x)$. It is necessary to construct membership function $\mu_1(x)$ which will explicitly define the left boundary $\mu_2(x)$, or to construct the left boundary of membership function $\mu_2(x)$ with which we define $\mu_1(x)$ explicitly.

Let us construct membership function $\mu_1(x)$. Let right boundary $\mu_2(x)$ look like $y = 1 - a_3 x - b_3$, $(-b_3 / a_3 > 0)$. Let us consider the objects allocated by an expert to level X_1. Having made paired comparisons of these objects and subsequent buildings similar to those above, we obtain linear function $y = a_1 x + b_1$, which is the right boundary of required function. If this function satisfies to two conditions: $y(0) > 1$; $y(-b_3 / a_3) < 0$, then

$$\mu_1(x) = \begin{cases} 1, & 0 \leq x \leq \dfrac{1-b_1}{a_1}; \\ a_1 x + b_1, & \dfrac{1-b_1}{a_1} < x \leq -\dfrac{b_1}{a_1}; \\ 0, & -\dfrac{b_1}{a_1} < x \leq 1, \end{cases}$$

$$\mu_2(x) = \begin{cases} 0, & 0 \leq x \leq \dfrac{1-b_1}{a_1}; \\ 1 - a_1 x - b_1, & \dfrac{1-b_1}{a_1} < x \leq -\dfrac{b_1}{a_1}; \\ 1, & -\dfrac{b_1}{a_1} < x \leq -\dfrac{b_3}{a_3}; \\ 1 - a_3 x - b_3, & -\dfrac{b_3}{a_3} < x \leq \dfrac{1-b_3}{a_3}; \\ 0, & \dfrac{1-b_3}{a_3} < x \leq 1. \end{cases}$$

If function $y = a_1 x + b_1$ satisfies the conditions: $y(0) > 1$; $y(-b_3 / a_3) \geq 0$, membership function $\mu_1(x)$ is defined under the condition $y(-b_3 / a_3) = -a_1(b_3 / a_3) + b_1 = 0$. In this case one of coefficients still remains unknown, it is determined from a normal equation. We obtain the membership functions

$$\mu_1(x)=\begin{cases}1, & 0\le x\le\dfrac{1-b_1}{a_1};\\[2mm]a_1x+a_1\dfrac{b_3}{a_3}, & \dfrac{1-b_1}{a_1}<x\le-\dfrac{b_3}{a_3};\\[2mm]0, & -\dfrac{b_3}{a_3}<x\le1;\end{cases}$$

$$\mu_2(x)=\begin{cases}0, & 0\le x\le\dfrac{1-b_1}{a_1};\\[2mm]1-a_1x-a_1\dfrac{b_3}{a_3}, & \dfrac{1-b_1}{a_1}<x\le-\dfrac{b_3}{a_3};\\[2mm]1-a_3x-b_3, & -\dfrac{b_3}{a_3}<x\le\dfrac{1-b_3}{a_3};\\[2mm]0, & \dfrac{1-b_3}{a_3}<x\le1.\end{cases}$$

If function $y=a_1x+b_1$ satisfies the conditions: $y(0)\le1$; $y(-b_3/a_3)<0$, it is supposed that $b_1=1$, and a_1 is obtained from a normal equation of a corresponding optimization problem. Membership functions look like

$$\mu_1(x)=\begin{cases}a_1x+1, & 0\le x\le-\dfrac{1}{a_1};\\[2mm]0, & -\dfrac{1}{a_1}<x\le1;\end{cases}$$

$$\mu_2(x)=\begin{cases}-a_1x, & 0\le x\le-\dfrac{1}{a_1};\\[2mm]1, & -\dfrac{1}{a_1}<x\le-\dfrac{b_3}{a_3};\\[2mm]1-a_3x-b_3, & -\dfrac{b_3}{a_3}<x\le\dfrac{1-b_3}{a_3};\\[2mm]0, & \dfrac{1-b_3}{a_3}<x\le1.\end{cases}$$

If function $y=a_1x+b_1$ satisfies the conditions: $y(0)\le1$; $y(-b_3/a_3)\ge0$, it is supposed that $b_1=1$; $y(-b_3/a_3)=-a_1(b_3/a_3)+b_1=0$; $a_1=a_3/b_3$. Then

$$\mu_1(x) = \begin{cases} \dfrac{a_3}{b_3}x+1, & 0 \le x \le -\dfrac{b_3}{a_3}; \\ 0, & -\dfrac{b_3}{a_3} < x \le 1; \end{cases}$$

$$\mu_1(x) = \begin{cases} -\dfrac{a_3}{b_3}x, & 0 \le x \le -\dfrac{b_3}{a_3}; \\ 1-a_3x-b_3, & -\dfrac{b_3}{a_3} < x \le \dfrac{1-b_3}{a_3}; \\ 0, & \dfrac{1-b_3}{a_3} < x \le 1. \end{cases}$$

On constructing the COSS membership functions for each evaluation y_i, $i = \overline{1, j}$ it is possible to determine its degrees of membership to COSS terms and to put in correspondence that term (or that level of a verbal scale) membership degree of which exceeds 0.5. Existence of such term for each evaluation y_i is stipulated by characteristics of COSS membership functions. The related outcomes obtained can generally not coincide with outcomes of expert preliminary allocations of evaluations y_i to levels of a verbal scale.

Example 2.3. Model-building of COSS "knowledge of students". A teacher appraises knowledge of eight students for a certain with two points: the first x_i; $i = \overline{1,8}$ is a result of examination and can accept values "F" ("unsatisfactory"), "C" "satisfactory"), "B" (good") and "A" ("excellent"); the second, y_i, is a result of testing and can take discrete values $0 \le k/n \le 1$, where k is a number of properly performed tasks, n is a number of all tasks performed (Table 2.1).

Let us construct COSS "knowledge of students" with terms $X_1 = F$, $X_2 = C$, $X_3 = B$, $X_4 = A$.

Let us start constructing with membership function $\mu_4(x)$ of term X_4. As to level "A" knowledge of only two students are referred to, let us join to them outcomes of the students with level "B", and let us suggest the teacher who carried out an evaluation to make paired comparisons of knowledge of all four students. Having arranged serial numbers of students according to the obtained points y_i, we obtain a conditional ordinal series

No. 2, No. 1, No. 4, No. 3.

to which the numerical ordinal series corresponds

0.9; 0.8; 0.7; 0.6.

The teacher makes paired comparisons of knowledge of students from the conditional ordinal series using Saati scale. The following matrix of paired comparisons is obtained:

$$A_4 = \begin{pmatrix} 1 & 2 & 5 & 7 \\ 1/2 & 1 & 3 & 4 \\ 1/5 & 1/3 & 1 & 2 \\ 1/7 & 1/4 & 1/2 & 1 \end{pmatrix}.$$

Table 2.1 Results of students' knowledge appraising

No.	x_i	y_i
1	Excellent ("A")	0.8
2	Excellent ("A")	0.9
3	"good" ("B")	0.6
4	"good" ("B")	0.7
5	"satisfactory" ("C")	0.3
6	"satisfactory" ("C")	0.1
7	"satisfactory" ("C")	0.2
8	"unsatisfactory" ("E")	0

We determine eigenvalues of this matrix: $\lambda_1 = 4{,}022$; $\lambda_2 = -0{,}010$; $\lambda_3 = -0{,}006 + 0{,}0294i$; $\lambda_4 = -0{,}006 - 0{,}0294i$. Let us select a maximum eigenvalue and the corresponding eigenvector $\omega_4 = (0{,}859; 0{,}466; 0{,}180; 0{,}109)$. Let us build a line $y = 2{,}537x - 1{,}499$ by four points $(0.9; 0.859)$, $(0.8; 0.466)$, $(0.7; 0.180)$, $(0.6; 0.109)$ applying a method of least squares. Let us check a value $y(1) = 2{,}537 - 1{,}499 = 1{,}038$. Apparently, the condition $y(1) \geq 1$ is satisfied, that ensures existence at least one point $x \in [0,1]: \mu_4(x) = 1$. Thus, $\mu_4(x) \equiv (0{,}985; 1; 0{,}385; 0)$.

By means of membership function $\mu_4(x)$, it is possible to define explicitly the right boundary of membership function $\mu_3(x)$, i.e. with $0{,}59 \leq x < 0{,}985$ $\mu_3(x) = 1 - \mu_4(x) = 2{,}499 - 2{,}537x$.

Obviously, the further building of $\mu_3(x)$ is not possible as there are no students who have got a point "B", and whose points are $y_i < 0{,}59$. In this case we come to building of membership function $\mu_2(x)$ of term X_2. Let us consider outcomes of the pupils who have got points "C" and "E". Let us arrange them as per decreasing points y_i. We obtain the following conditional ordinal series

No.5, No. 7, No. 6, No. 8.

to which the numerical ordinal number corresponds

0.3; 0.2; 0.1; 0.

The teacher makes paired comparisons of knowledge of students from the conditional ordinal series using Saati scale. The following matrix of paired comparisons is obtained:

$$A_2 = \begin{pmatrix} 1 & 2 & 5 & 9 \\ \dfrac{1}{2} & 1 & 2 & 5 \\ \dfrac{1}{5} & \dfrac{1}{2} & 1 & 2 \\ \dfrac{1}{9} & \dfrac{1}{5} & \dfrac{1}{2} & 1 \end{pmatrix}.$$

We determine eigenvalues of this matrix: $\lambda_1 = 4{,}008$; $\lambda_2 = -0{,}004$; $\lambda_3 = -0{,}001 + 0{,}175i$; $\lambda_4 = -0{,}001 - 0{,}175i$. Let us select a maximum eigenvalue and the corresponding eigenvector $\omega_2 = (0{,}879; 0{,}427; 0{,}191; 0{,}093)$. Let us build a line $y = 2{,}6x - 0{,}01$ by four points $(0.3; 0.879)$, $(0.2; 0.427)$, $(0.1; 0.191)$, $(0; 0.093)$, applying a method of least squares. Let us check the value $y(0{,}59) = 1{,}534 - 0{,}01 = 1{,}524$. With $y = 0$, $x = 0{,}0038$; with $y = 1$, $x = 0{,}39$.

Thus, the left and right boundaries of membership functions $\mu_2(x)$ and $\mu_3(x)$ are defined. To complete constructing the functions, let us find the equation of a line $y = 2{,}95 - 5x$, which is passing through two points $(0.39; 1)$, $(0.59; 0)$. This line will limit membership function $\mu_2(x)$ on the right, and the line $y = 1 - 2{,}95 + 5x = 5x - 1{,}95$ will limit membership function $\mu_3(x)$ on the left. Thus,

$$\mu_3(x) = (0{,}59; 0{,}2; 0{,}385); \quad \mu_2(x) = (0{,}39; 0{,}3862; 0{,}2).$$

Membership function $\mu_1(x)$ is defined using function $\mu_2(x)$ in the following manner:

$$\mu_1(x) = (0; 0{,}0038; 0; 0{,}3862)$$

2.4 Formalization of Linguistic Values of Characteristics on the Basis of Direct Inquiry of a Single Expert Regarding Typical Representatives

By processing and analyzing the information which can have both qualitative and quantitative nature, a person makes different decisions. Quantitative characteristics are those which can be measured. For example, enterprise profits and losses, age of a person, object weight etc. could be treated as quantitative ones. Qualitative characteristics cannot be measured in terms of quantities. For example, some of those characteristics are human appearance, human knowledge, success of performance of professional duties etc.

Thus, characteristics are mainly described with different "languages": as a rule, for quantitative characteristics we use usual numbers which are referred to as physical values of characteristics; for qualitative ones they are words of a natural language which are referred to as linguistic values. Linguistic values can be defined also for quantitative characteristics; however, for the qualitative characteristics physical values cannot be defined.

Two model-building methods of qualitative characteristics COSS are described above. Formalization of the information based on the expert evaluations of appearances of different qualitative characteristics allows operating not with values of the characteristics measured in different scales, but with membership functions of the concepts applied to an evaluation of characteristics of real objects.

These methods are inapplicable to construct quantitative characteristic COSS. So, we offer to construct membership functions of COSS terms on the basis of direct inquiry of a single expert, these COSS's can be applied to formalize both quantitative and qualitative characteristics pari passu.

It is worth mentioning that a COSS based on inquiry of experts will always possess some property of uniqueness, i.e. it reflects judgments of the experts who often use the information known to few people who are "in gathering".

Actually, if one wants to build COSS "height" = {low, average, high, very high} from point of view of Moscow and Tokyo experts, then, obviously, there will be two spaces with a different collection of membership functions. If one wants to build COSS "profit" = {very low, low, average, high, very high}, the money equivalent which is considered as high profit, will dramatically differ for different firms. It is just the case when defining similar categories experts use the information known to few people only, on the one hand, and unique for a certain firm, on the other hand.

Model-building techniques of COSS term-sets membership functions for quantitative and qualitative characteristics have in essence identical approaches and differ only in universal sets.

Let us construct COSS with a title X and term-set $T(X) = \{X_1, X_2, ..., X_m\}$, where X_1, X_m are terms corresponding to the minimum and maximum intensity degree of characteristic appearance within a universal set $U = [a, b]$.

Let us assume that an expert defines typical intervals $\left(x_l^1, x_l^2\right)$ for terms X_l; $l = \overline{1, m}$, and these intervals are equal to unity for all points of membership function of corresponding terms. For some terms, points (one for each term) can be typical rather than intervals. Without loss of generality, typical intervals for qualitative characteristics can be defined by the following procedure. Let there be a test for demonstration of studied qualitative characteristic for the object (0 is a minimum quantity of points which an object can obtain as testing results, i.e. complete absence of characteristic appearance; n is maximum quantity of points, i.e. its total availability of the characteristic appearance). Let us normalize all possible test points with a maximum point n and let us give the chance to the expert to answer regarding typical intervals for terms X_l belonging to [0, 1].

The questions asked to the expert can be formulated as follows: "What interval of normalized test points you consider typical for a term?". In case of quantitative characteristic the expert defines subsets of universal set which are typical for each of terms from his/her point of view.

Let us assume that the fuzzy numbers corresponding to COSS terms are $\wedge$-numbers; and for functions $L(x)$, $R(x)$ the side condition (1 *) is satisfied (see paragraph 2.2]. Membership functions of terms X_i:

with even m

$$\mu_1(x) = \begin{cases} 1, & a \leq x \leq x_1^2; \\ L\left(\dfrac{x - x_1^2}{x_2^1 - x_1^2}\right), & x_1^2 < x \leq x_2^1; \\ 0, & x_2^1 < x \leq b; \end{cases}$$

$$\mu_2(x) = \begin{cases} 0, & a \leq x \leq x_1^2; \\ L\left(\dfrac{x_2^1 - x}{x_2^1 - x_1^2}\right), & x_1^2 < x \leq x_2^1; \\ 1, & x_2^1 < x \leq x_2^2; \\ R\left(\dfrac{x - x_2^2}{x_3^1 - x_2^2}\right), & x_2^2 < x \leq x_3^1; \\ 0, & x_3^1 < x \leq b; \end{cases}$$

$$\mu_{m-1}(x) = \begin{cases} 0, & a \leq x \leq x_{m-2}^2; \\ R\left(\dfrac{x_{m-1}^1 - x}{x_{m-1}^1 - x_{m-2}^2}\right), & x_{m-2}^2 < x \leq x_{m-1}^1; \\ 1, & x_{m-1}^1 < x \leq x_{m-1}^2; \\ L\left(\dfrac{x - x_{m-1}^2}{x_m^1 - x_{m-1}^2}\right), & x_{m-1}^2 < x \leq x_m^1; \\ 0, & x_m^1 < x \leq b; \end{cases}$$

$$\mu_m(x) = \begin{cases} 0, & a \leq x \leq x_{m-1}^2; \\ L\left(\dfrac{x_m^1 - x}{x_m^1 - x_{m-1}^2}\right), & x_{m-1}^2 < x \leq x_m^1; \\ 1, & x_m^1 < x \leq b \end{cases}$$

With odd m two last membership functions $L(x)$, $R(x)$ interchange their positions.

Constructed COSS reflects knowledge and subjective judgment of the expert taking part in inquiry.

If deriving of the additional information on the values of membership functions in the points of universal set which lies between typical intervals of the adjacent terms is possible, then it is also possible to clarify a form of these functions.

Let us consider typical intervals $\left(x_l^1, x_l^2\right)$ and $\left(x_{l+1}^1, x_{l+1}^2\right)$ of the adjacent terms X_l and X_{l+1}, $l = \overline{1, m-1}$. If one assumes that the right and left boundaries of membership function of the fuzzy number, which correspond to terms X_l and X_{l+1} are linear functions, then they have the following explicit analytical form:

$$y = 1 - \frac{x - a_l^2}{a_{l+1}^1 - a_l^2} \quad \text{and} \quad y = 1 + \frac{x - a_{l+1}^1}{a_{l+1}^1 - a_l^2}$$

Let $b \in \left(x_l^2, x_{l+1}^1\right)$. If the expert agrees that

$$\mu_{l+1}(b) = 1 + \frac{b - a_{l+1}^1}{a_{l+1}^1 - a_l^2},$$

then boundaries are to be selected as linear functions. If the expert considers that

$$\mu_{l+1}(b) > 1 + \frac{b - a_{l+1}^1}{a_{l+1}^1 - a_l^2},$$

then boundaries are to be selected as the functions shown in Fig. 2.5.a. If the expert considers that

$$\mu_{l+1}(b) < 1 + \frac{b - a_{l+1}^1}{a_{l+1}^1 - a_l^2},$$

then boundaries are to be selected as the functions shown in Fig. 2.5.b.

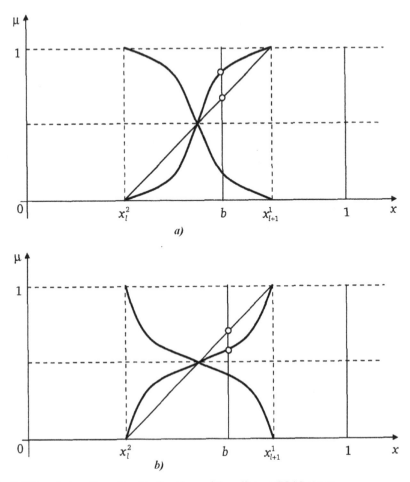

Fig. 2.5 Boundaries of membership functions of the adjacent COSS terms:

$$\mathbf{a} - \mu_{l+1}(b) > 1 + \frac{b - a_{l+1}^1}{a_{l+1}^1 + a_l^2} \; ; \mathbf{b} - \mu_{l+1}(b) < 1 + \frac{b - a_{l+1}^1}{a_{l+1}^1 - a_l^2}$$

Example 2.4. Model-building of COSS "net current assets/turnover assets ratio". Data on audit and analysis of enterprises of machine-building and food industries, and of telecommunications facilities production enterprises was used. The net current assets/turnover assets ratio was studied for each enterprise. Based on the analysis carried out, let us construct COSS "net current assets/turnover assets ratio" using terms "very low", "low", "average", "high", "very high". The experts who surveyed the enterprises were provided with intervals of U ratio values, which are typical for every term $U = [-1,1]$: $[b_{11}, b_{12}] = [-1; -0,005]$; $[b_{21}, b_{22}] = [0; 0,09]$, $[b_{31}, b_{32}] = [0,11; 0,3]$, $[b_{41}, b_{42}] = [0,35; 0,45]$, $[b_{51}, b_{52}] = [0,5; 1]$.

Then, membership functions $\mu_i(x)$, $i=\overline{1,5}$ for terms "very low", "low", "average", "high", "very high" are T-numbers and look like: $\mu_1(x)=(-1;-0,005;0;0,005)$; $\mu_2(x)=(0;0,09;0,005;0,02)$; $\mu_3(x)=(0,11;0,3;0,02;0,05)$; $\mu_4(x)=(0,35;0,45;0,05;0,05)$; $\mu_5(x)=(0,5;1;0;0,5;0)$, accordingly.

2.5 Formalization of Linguistic Values of Characteristics on the Basis of Direct Inquiry of a Single Expert Regarding a Partition of Universal Set

Let, for any reasons, an expert has difficulties in defining typical intervals for COSS terms, but he/she can divide universal set into disjoint intervals each mapped to one of terms and being a set of 0.5-level of the fuzzy number corresponding to this term. Let us assume that the fuzzy numbers corresponding to COSS terms are Λ-numbers, and the side condition (1*) is satisfied for functions $L(x)$, $R(x)$. Let us denote membership functions of terms X_l as $\mu_l(x)$, $l=\overline{1,m}$. Let us denote length of the interval corresponding to term X_l through

$$c_l, \quad \sum_{l=1}^{m} c_l = b - a; U = [a,b].$$

Let us construct membership functions $\mu_l(x)$ as curvilinear trapezoids with midlines equal to c_l

1. If $c_m \le c_{m-1}$, then

$$\mu_m(x) = \begin{cases} 0, & a \le x \le \left(b - \dfrac{3c_m}{2}\right); \\[2ex] L\left(\dfrac{b - \dfrac{c_m}{2} - x}{c_m}\right), & \left(b - \dfrac{3c_m}{2}\right) < x \le \left(b - \dfrac{c_m}{2}\right); \\[2ex] 1, & \left(b - \dfrac{c_m}{2}\right) < x \le b. \end{cases}$$

2. If $c_m > c_{m-1}$, then

$$\mu_m(x) = \begin{cases} 0, & a \le x \le \left(b - c_m - \dfrac{c_{m-1}}{2}\right); \\[3mm] L\left(\dfrac{b - c_m + \dfrac{c_{m-1}}{2} - x}{c_{m-1}}\right), & \left(b - c_m - \dfrac{c_{m-1}}{2}\right) < x \le \left(b - c_m + \dfrac{c_{m-1}}{2}\right); \\[3mm] 1, & \left(1 - c_m + \dfrac{c_{m-1}}{2}\right) < x \le b. \end{cases}$$

Let us construct membership function of term X_{m-1}:

1. If $c_{m-1} \ge \max(c_m, c_{m-2})$, then

$$\mu_{m-1}(x) = \begin{cases} 0, & a \le x \le \left(\sum\limits_{l=1}^{m-3} c_l + \dfrac{c_{m-2}}{2}\right); \\[4mm] R\left(\dfrac{\sum\limits_{l=1}^{m-2} c_l + \dfrac{c_{m-2}}{2} - x}{c_{m-2}}\right), & \left(\sum\limits_{l=1}^{m-3} c_l + \dfrac{c_{m-2}}{2}\right) < x \le \left(\sum\limits_{l=1}^{m-2} c_l + \dfrac{c_{m-2}}{2}\right); \\[4mm] 1, & \left(\sum\limits_{l=1}^{m-2} c_l + \dfrac{c_{m-2}}{2}\right) < x \le \left(b - \dfrac{3c_m}{2}\right); \\[4mm] L\left(\dfrac{x - b + \dfrac{3c_m}{2}}{c_m}\right), & \left(1 - \dfrac{3c_m}{2}\right) < x \le \left(b - \dfrac{c_m}{2}\right); \\[4mm] 0, & \left(b - \dfrac{c_m}{2}\right) < x \le b. \end{cases}$$

2. If $c_m < c_{m-1} < c_{m-2}$, then

$$\mu_{m-1}(x) = \begin{cases} 0, & a \le x \le \left(1 - c_m - \dfrac{3c_{m-1}}{2}\right); \\[2mm] R\left(\dfrac{b - c_m - \dfrac{c_{m-1}}{2} - x}{c_{m-1}}\right), & \left(b - c_m - \dfrac{3c_{m-1}}{2}\right) < x \le \left(b - c_m - \dfrac{c_{m-1}}{2}\right); \\[2mm] 1, & \left(b - c_m - \dfrac{c_{m-1}}{2}\right) < x \le \left(b - \dfrac{3c_m}{2}\right); \\[2mm] L\left(\dfrac{x - b + \dfrac{3c_m}{2}}{c_m}\right), & \left(b - \dfrac{3c_m}{2}\right) < x \le \left(b - \dfrac{c_m}{2}\right); \\[2mm] 0, & \left(b - \dfrac{c_m}{2}\right) < x \le b. \end{cases}$$

3. If $c_{m-2} < c_{m-1} < c_m$, then

$$\mu_{m-1}(x) = \begin{cases} 0, & a \le x \le \left(\sum_{l=1}^{m-3} c_l + \dfrac{c_{m-2}}{2}\right); \\[2mm] R\left(\dfrac{\sum_{l=1}^{m-2} c_l + \dfrac{c_{m-2}}{2} - x}{c_{m-2}}\right), & \left(\sum_{l=1}^{m-3} c_l + \dfrac{c_{m-2}}{2}\right) < x \le \left(\sum_{l=1}^{m-2} c_l + \dfrac{c_{m-2}}{2}\right); \\[2mm] 1, & \left(\sum_{l=1}^{m-2} c_l + \dfrac{c_{m-2}}{2}\right) < x \le \left(b - c_m - \dfrac{c_{m-1}}{2}\right); \\[2mm] L\left(\dfrac{x - b + c_m + \dfrac{c_{m-1}}{2}}{c_{m-1}}\right), & \left(b - c_m - \dfrac{c_{m-1}}{2}\right) < x \le \left(b - c_m + \dfrac{c_{m-1}}{2}\right); \\[2mm] 0, & \left(b - c_m + \dfrac{c_{m-1}}{2}\right) < x \le b. \end{cases}$$

4. If $c_{m-1} \le \min(c_m, c_{m-2})$, then

$$\mu_{m-1}(x) = \begin{cases} 0, & a \le x \le \left(b - c_m - \dfrac{3c_{m-1}}{2}\right); \\[2mm] R\left(\dfrac{b - c_m - \dfrac{c_{m-1}}{2} - x}{c_{m-1}}\right), & \left(b - c_m - \dfrac{3c_{m-1}}{2}\right) < x \le \left(b - c_m - \dfrac{c_{m-1}}{2}\right); \\[2mm] L\left(\dfrac{x - b + c_m + \dfrac{c_{m-1}}{2}}{c_{m-1}}\right), & \left(b - c_m - \dfrac{c_{m-1}}{2}\right) < x \le \left(b - c_m + \dfrac{c_{m-1}}{2}\right); \\[2mm] 0, & \left(b - c_m + \dfrac{c_{m-1}}{2}\right) < x \le b. \end{cases}$$

Similarly to $\mu_{m-1}(x)$ membership functions $\mu_l(x)$, $l = \overline{2, m-2}$ for term X_1 are constructed.

With even number of terms:

1. If $c_1 \le c_2$, then

$$\mu_1(x) = \begin{cases} 1, & a \le x \le \dfrac{c_1}{2}; \\[2mm] L\left(\dfrac{x - \dfrac{c_1}{2}}{c_1}\right), & \dfrac{c_1}{2} < x \le \dfrac{3c_1}{2}; \\[2mm] 0, & \dfrac{3c_1}{2} < x \le b. \end{cases}$$

2. If $c_1 > c_2$, then

$$\mu_1(x) = \begin{cases} 1, & a \le x \le \left(c_1 - \dfrac{c_2}{2}\right); \\[2mm] L\left(\dfrac{x - c_1 + \dfrac{c_2}{2}}{c_2}\right), & \left(c_1 - \dfrac{c_2}{2}\right) < x \le \left(c_1 + \dfrac{c_2}{2}\right); \\[2mm] 0, & \left(c_1 + \dfrac{c_2}{2}\right) < x \le b. \end{cases}$$

With odd number of terms:

1. If $c_1 \le c_2$, then

$$\mu_1(x) = \begin{cases} 1, & a \le x \le \dfrac{c_1}{2}; \\[2ex] R\left(\dfrac{x - \dfrac{c_1}{2}}{c_1}\right), & \dfrac{c_1}{2} < x \le \dfrac{3c_1}{2}; \\[2ex] 0, & \dfrac{3c_1}{2} < x \le b. \end{cases}$$

2. If $c_1 > c_2$, then

$$\mu_1(x) = \begin{cases} 1, & a \le x \le \left(c_1 - \dfrac{c_2}{2}\right); \\[2ex] R\left(\dfrac{x - c_1 + \dfrac{c_2}{2}}{c_2}\right), & \left(c_1 - \dfrac{c_2}{2}\right) < x \le \left(c_1 + \dfrac{c_2}{2}\right); \\[2ex] 0, & \left(c_1 + \dfrac{c_2}{2}\right) < x \le b. \end{cases}$$

If additional information on values of membership functions in the points of universal set, which lie between typical intervals of the adjacent terms is available, the updating of the form of membership functions is made in the same manner as the previous method shows.

2.6 Formalization of Linguistic Values of Properties on the Basis of Direct Inquiry of Expert Groups about Regarding Representatives

Let k experts offer intervals of values, which are typical for each of m terms from their point of view. It is possible to represent outcomes of their evaluations in the form of the matrix

$$A = \begin{pmatrix} \left(x_{11}^1, x_{11}^2\right) & \left(x_{12}^1, x_{12}^2\right) & \cdots & \left(x_{1m}^1, x_{1m}^2\right) \\ \left(x_{21}^1, x_{21}^2\right) & \left(x_{22}^1, x_{22}^2\right) & \cdots & \left(x_{2m}^1, x_{2m}^2\right) \\ \cdot & \cdot & \cdot & \cdot \\ \left(x_{k1}^1, x_{k1}^2\right) & \left(x_{k2}^1, x_{k2}^2\right) & \cdots & \left(x_{km}^1, x_{km}^2\right) \end{pmatrix}.$$

Let

$$\left(0,x_1^2\right)=\bigcap_{i=1}^{k}\left(x_{i1}^1,x_{i1}^2\right);$$

$$\left(x_2^1,x_2^2\right)=\bigcap_{i=1}^{k}\left(x_{i2}^1,x_{i2}^2\right);$$

. . . .

$$\left(x_m^1,1\right)=\bigcap_{i=1}^{k}\left(x_{im}^1,x_{im}^2\right).$$

If

$$[0,1]=\bigcap_{l=1}^{m}\left(x_l^1,x_l^2\right),$$

then COSS term-set membership functions are transformed to characteristic functions of intervals $\left(x_l^1,x_l^2\right)$, $l=\overline{1,m}$. If one of intervals $\left(x_l^1,x_l^2\right)=\varnothing$, then instead of intersection operation, r-composition of intervals operation [124] is applied. Operation of r-composition of intervals is introduced through union and intersection operations as follows:

$$\left(\begin{array}{c}k\\r\\i=1\end{array}\right)\left(x_i^1,x_i^2\right)=\bigcup_{i_1\neq\ldots\neq i_{k-r+1}}\left[\left(x_{i_1}^1,x_{i_1}^2\right)\bigcap\ldots\bigcap\left(x_{i_{k-r+1}}^1,x_{i_{k-r+1}}^2\right)\right];$$

$$\left(\begin{array}{c}k\\r\\i=1\end{array}\right)\left(x_i^1,x_i^2\right)=\bigcap_{i_1\neq\ldots\neq i_r}\left[\left(x_{i_1}^1,x_{i_1}^2\right)\bigcup\ldots\bigcup\left(x_{i_r}^1,x_{i_r}^2\right)\right].$$

The selection of $r\approx k/2$ is obvious because the "most representative" expert should prefer an evaluation, which is remote from extreme evaluations and occupying "middle" position. If $\exists l:\left(x_l^1,x_l^2\right)\bigcap\left(x_{l-1}^1,x_{l-1}^2\right)=\left(a_l,b_l\right)\neq\phi$, or $\left(x_l^1,x_l^2\right)\bigcap\left(x_{l+1}^1,x_{l+1}^2\right)=\left(a_l,b_l\right)\neq\varnothing$, then instead of intervals $\left(x_{l-1}^1,x_{l-1}^2\right)$, $\left(x_l^1,x_l^2\right)$, $\left(x_{l+1}^1,x_{l+1}^2\right)$ the following intervals are considered, accordingly

$$\left(\tilde{x}_{l-1}^1,\tilde{x}_{l-1}^2\right)=\left(x_{l-1}^1,x_{l-1}^2\right)\backslash\left(a_l,b_l\right);\ \left(\tilde{x}_l^1,\tilde{x}_l^2\right)=\left(x_l^1,x_l^2\right)\backslash\left(a_l,b_l\right)$$

$$\left(\tilde{x}_{l+1}^1,\tilde{x}_{l+1}^2\right)=\left(x_{l+1}^1,x_{l+1}^2\right)\backslash\left(a_l,b_l\right).$$

Let us assume that intervals $\left(x_l^1,x_l^2\right)$ and $U=[a,b]$ typical for COSS terms are defined. Let us consider that the fuzzy numbers corresponding to COSS terms are Λ-numbers, and the side condition (1*) is satisfied for functions $L(x)$, $R(x)$. Let us denote membership functions of terms X_l with $\mu_l(x)$.

Then with even m

$$\mu_1(x) = \begin{cases} 1, & a \le x \le x_1^2; \\ L\left(\dfrac{x - x_1^2}{x_2^1 - x_1^2}\right), & x_1^2 < x \le x_2^1; \\ 0, & x_2^1 < x \le b; \end{cases}$$

$$\mu_2(x) = \begin{cases} 0, & a \le x \le x_1^2; \\ L\left(\dfrac{x_2^1 - x}{x_2^1 - x_1^2}\right), & x_1^2 < x \le x_2^1; \\ 1, & x_2^1 < x \le x_2^2; \\ R\left(\dfrac{x - x_2^2}{x_3^1 - x_2^2}\right), & x_2^2 < x \le x_3^1; \\ 0, & x_3^1 < x \le b; \end{cases}$$

$$\cdots\cdots\cdots\cdots\cdots\cdots\cdots\cdots\cdots\cdots\cdots$$

$$\mu_{m-1}(x) = \begin{cases} 0, & a \le x \le x_{m-2}^2; \\ R\left(\dfrac{x_{m-1}^1 - x}{x_{m-1}^1 - x_{m-2}^2}\right), & x_{m-2}^2 < x \le x_{m-1}^1; \\ 1, & x_{m-1}^1 < x \le x_{m-1}^2; \\ L\left(\dfrac{x - x_{m-1}^2}{x_m^1 - x_{m-1}^2}\right), & x_{m-1}^2 < x \le x_m^1; \\ 0, & x_m^1 < x \le b; \end{cases}$$

$$\mu_m(x) = \begin{cases} 0, & a \le x \le x_{m-1}^2; \\ L\left(\dfrac{x_m^1 - x}{x_m^1 - x_{m-1}^2}\right), & x_{m-1}^2 < x \le x_m^1;. \\ 1, & x_m^1 < x \le b \end{cases}$$

With odd m two last membership functions $L(x)$, $R(x)$ interchange.

If obtaining of an additional information on values of membership functions in the points of universal set, which lie between typical intervals of the adjacent terms is possible, the form of membership functions of terms is updated.

Example 2.5. Model-building of COSS "pressure at the high-pressure preheater inlet". Let us construct COSS "pressure at the high-pressure preheater inlet" with terms "very low pressure", "low pressure", "normal pressure", "high pressure".

Three experts offer their typical values for each of terms. The first expert: {1.1}, {1.7}, {4}, {6.7}. The second expert: [1.1; 1.3], {1.7}, {4}, [6.6; 6.7]. The third expert: {1.1}, [1.6; 1.7], {4}, {6.7}. Based on intersections of typical values corresponding to each of terms, we obtain COSS which membership functions are shown in Fig. 2.6.

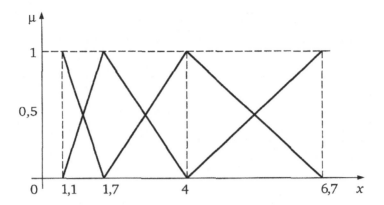

Fig. 2.6 Membership functions of COSS terms "pressure at the high-pressure preheater inlet"

2.7   Analogues of Errors of the First and Second Kinds and A Reliability Indicator for Models of Expert Evaluations of Characteristics

In [28] the quantity indicator of COSS quality is defined, namely, fuzziness degree. Let us find analogues of errors of the first and second kinds and reliability indicator for models of expert evaluations of characteristics.

Let $X = \{\mu_l(x);\ l = \overline{1,m}\}$ be COSS; $\mu_l(x) \equiv (a_1^l, a_2^l, a_L^l, a_R^l)$, $U = [a,b]$.

For membership functions $\mu_l(x)$, $\mu_{l+1}(x)$; $l = \overline{1,m-1}$ of the adjacent COSS terms there are uncertainty zones under $[x \in U : 0 < \mu_l(x) < 1; 0 < \mu_{l+1}(x) < 1]$ which potency is equal to a_R^l. Based on characteristics of COSS membership functions and concept of a probabilistic geometry, it is possible to consider $\alpha_l = \beta_l = a_R^l / 2(b-a)$ as analogues of errors of the first and second kinds for X_l (analogues of errors of the second and first kinds for X_{l+1}, accordingly). Then analogues of errors of the first and second kinds for models of expert evaluations of characteristics will be referred to as

$$\alpha = \beta = \max_{l=1,m-1}\{\alpha_l\}.$$

and analogue of a reliability indicator

$$V = (1-\alpha)^2.$$

$$V = \left(1 - \max_{l=1,m-1}\left\{\frac{a_R^l}{2(b-a)}\right\}\right)^2.$$

is analogue of a reliability indicator for models of expert evaluations of characteristics $X = \{\mu_l(x);\ l = \overline{1,m}\}$; $\mu_l(x) \equiv (a_1^l, a_2^l, a_L^l, a_R^l)$, defined over universal set $[a,b]$.

The analogue of a reliability indicator is connected with fuzziness degree as follows (with $f(x)=1-x$): if fuzziness degree of the model is minimum, i.e. is equal to zero, the analogue of a reliability indicator is maximum and equal to unity; if fuzziness degree is maximum, i.e. equal to 0.5, the analogue of a reliability indicator is minimum and equal to 0.25. The certain indicator allows essential expanding the information obtained on the basis of fuzziness degree because with identical fuzziness degree the models of expert evaluation of characteristics can have different values of reliability indicator analogue.

Example 2.6. Definition of reliability indicator analogue for models of expert evaluations of characteristics. Let us consider models of an expert evaluation of production quality from the example 2.1 and knowledge of students from the example 2.2. For model of production quality evaluation, analogues of errors of the first and second kinds are equal to 0.1482, and analogue of a reliability indicator is equal to 0.7255. For model of evaluation of students' knowledge, analogues of errors of the first and second kinds are equal to 0.15, and the analogue of a reliability indicator is equal to 0.7225.

2.8 Examples of Application of Complete Orthogonal Semantic Spaces in Problems of Information Analysis and Decision Making

Example 2.7. A multicriterion selection of software. [125—127] The current software market offers a great many of products for which quality evaluation systems of characteristics are developed. Complexity of a software selection is explained by a number of the objective and subjective reasons considered in details in [128—133].

One of such reasons is use of quantitative and qualitative characteristics, to measure which the various scales are applied: numerical, ordinal, verbal, etc. At that, some characteristics can be provided as values of a membership to levels of linguistic (verbal) scales.

Another reason is discrepancy of quality characteristics, which leads to a selection ambiguity and makes additive convolution of comparable indicators spurious.

Let us show a solution of software multicriterion selection problem on the basis of fuzzy conclusion rules.

Let us consider the following characteristics of software quality: modifiability, studiability, completeness. Let us add software price because it is of interest for a user along with quality characteristics.

Modifiability is a characteristic which simplifies introducing of necessary modifications and updating and includes concepts of expansibility, structuredness and modularity.

Studiability is a characteristic which allows minimisation of efforts in studying and understanding software programs and documentation and includes informativeness, clearness, structuredness and readability.

Within the scope of four formulated characteristics, one of three software products conventionally named as "A", "B" and "C" is selected to analyze financial standing of an enterprise.

For an evaluation of their characteristics the following scales are used:

- Price — "low", "mean", "high";
- Modifiability — "low", "mean", "high";
- Studiability — "low", "mean", "high";
- Completeness — "incompleteness", "partial completeness", "basic completeness", "complete completeness".

Based on the results of the software manufacturer price policy study, COSS was constructed with membership functions of terms: $\mu_1^1(x) \equiv (2000, 2500, 0; 500)$; $\mu_2^1(x) \equiv (3000; 3500; 500; 500)$; $\mu_3^1(x) \equiv (4000; 5000; 500; 0)$.   Membership functions of COSS terms "modifiability" are denoted by $\mu_1^2(x)$, $\mu_2^2(x)$, $\mu_3^2(x)$, accordingly, terms "studiability" - by $\mu_1^3(x)$, $\mu_2^3(x)$, $\mu_3^3(x)$, accordingly, and terms "completeness" - by $\mu_1^4(x)$, $\mu_2^4(x)$, $\mu_3^4(x)$, accordingly.

Customer survey has yielded following outcomes: if "low price, high modifiability, high studiability, and completeness available" or "low price, high modifiability, high studiability, and basic completeness available", or "low price, mean modifiability, high studiability, and completeness available" or "low price, high modifiability, mean studiability, and completeness available" or "low price, high modifiability, high studiability, and completeness available", or "low price, high modifiability, high studiability, and basic completeness available", or "low price, high modifiability, mean studiability and completeness available", or "low price, mean modifiability, high studiability and completeness available", the software is thought proper for purchase.

Eight fuzzy selection (conclusion) rules are included in the customer formulated preference. In total, there are 108 such rules as four characteristics are considered, one of which has four linguistic values, and each of three other has three values.

Outcomes of system from eight selection rules are degrees of considered software membership to fuzzy set "the software which is thought proper for purchase". When operating with COSS terms membership functions, the conjunctions "and" and "or" are treated as operations "min" and "max", accordingly.

Experts have estimated software products "A", "B", "C" and have obtained the following outcomes:

$\mu_1^1(A) = 0,68$; $\mu_2^1(A) = 0,32$; $\mu_1^1(B) = 0,74$; $\mu_2^1(B) = 0,26$; $\mu_1^1(C) = 0,46$; $\mu_2^1(C) = 0,54$; $\mu_2^2(A) = 0,6$; $\mu_3^2(A) = 0,4$; $\mu_2^2(B) = 0,7$; $\mu_3^2(B) = 0,3$; $\mu_2^2(C) = 0,1$; $\mu_3^2(C) = 0,9$; $\mu_2^3(A) = 0,4$; $\mu_3^3(A) = 0,6$; $\mu_2^3(B) = 0,1$;

$$\mu_3^3(B)=0{,}9 \; , \; \mu_2^3(C)=0{,}3 \; , \; \mu_3^3(C)=0{,}7 \; , \; \mu_3^4(A)=0{,}3 \; , \; \mu_4^4(A)=0{,}7$$
$$\mu_3^4(B)=0{,}2 \; , \; \mu_4^4(A)=0{,}8 \; , \; \mu_3^4(C)=0{,}2 \; , \; \mu_4^4(C)=0{,}8 \; .$$

To define the "A", "B", "C" values of membership to fuzzy set "the software which is thought proper for purchase" with membership function, the following rule was applied [15]:

$$\mu(x)=\max \begin{bmatrix} \left(\min\!\left[\mu_1^1(x),\mu_3^2(x),\mu_3^3(x),\mu_4^4(x)\right]\!\!;\min\!\left[\mu_1^1(x),\mu_3^2(x),\mu_3^3(x),\mu_4^4(x)\right]\!\right) \\ \min\!\left[\mu_1^1(x),\mu_3^2(x),\mu_3^3(x),\mu_4^4(x)\right]\!\!;\min\!\left[\mu_1^1(x),\mu_2^2(x),\mu_3^3(x),\mu_4^4(x)\right] \\ \min\!\left[\mu_2^1(x),\mu_3^2(x),\mu_3^3(x),\mu_4^4(x)\right]\!\!;\min\!\left[\mu_2^1(x),\mu_3^2(x),\mu_3^3(x),\mu_4^4(x)\right] \\ \left(\min\!\left[\mu_2^1(x),\mu_3^2(x),\mu_3^3(x),\mu_4^4(x)\right]\!\!;\min\!\left[\mu_2^1(x),\mu_3^2(x),\mu_3^3(x),\mu_4^4(x)\right]\!\right) \end{bmatrix}$$

The results of computations are as follows:

$$\mu(A)=0{,}6 \; , \; \mu(B)=0{,}7 \; , \; \mu(C)=0{,}54 \; .$$

On the basis of the obtained results the decision on software "B" purchase is made.

Example 2.8. Definition of analogy degree for products. While solving engineering problems, it is often necessary to define analogy degree of a product by some parameters a and b [35]

$$\rho_{ab} = 1 - \frac{|a-b|}{B-A}, \tag{2.4}$$

where $[A, B]$ is a range of parameter values.

Advantage of this formula is simplicity of calculations, and its disadvantage is that analogy degree only depends on a difference of parameter values and does not depend on location of these values over the whole area $[A, B]$.

This disadvantage is the reason of the fact that outcomes obtained with this formula do not always match expert experience. The matter is that a qualified expert selects some base values of parameter within a set of parameters and makes comparison of products on the basis of selected values. For example, an expert selects low, mean and high values, and defines analogy of products depending on what linguistic values of parameter its numerical values belong to [35] gives the example of definition of analogy degree of products referred as "high pressure preheater" by the parameter "inlet steam pressure". The parameter has a definition range [1.1, 6.7]. In this example analogy degrees of products with values (1.1 and 1.5) and (6.1 and 6.6) are defined. The above formula provides analogy degree 0.93 for first pair of products and 0.91 for second pair. In [35] it is stated that these outcomes do not match the experience of experts. Moreover, based on the analysis of expert experience in designing similar products, products of the first pair are much less similar to each other than products of the second pair. In [35] a new formula to define product analogy degree based on membership functions of term-sets "low pressure", "pressure close to 4", "high pressure" of semantic space "inlet steam pressure" is offered. However, a shortage of this formula is in lack of symmetry attribute, i.e. $P_{ab} \neq P_{ba}$ [134] contains a formula of product analogy degree determination based on COSS term-sets membership functions, which is free from this shortage.

Let us consider $U = [A, B]$ as a range of parameter values. Based on his/her experience, an expert selects m linguistic values of this parameter and specified corresponding typical numerical values for each linguistic value. Typical values can be specified by one number or by the entire interval. Depending upon that, membership functions of term-sets $\mu_l(x)$, $l = \overline{1, m}$ (see §2.4) are membership functions of unimodal or tolerance numbers from population Λ constructed in §2.1.

Let us identify $|\mu_l(a) - \mu_l(b)|$ as a measure of information loss for values a and b within the limits of l-th term-set. Let us determine an analogy degree of products with values a and b for the considered parameter by the formula

$$\rho_{ab} = 1 - \frac{\sum_{l=1}^{m} |\mu_l(a) - \mu_l(b)|}{2}.$$

From this formula it follows that if a and b belong to tolerance areas of one function $[\mu_l(a) = \mu_l(b) = 1]$, then $\rho_{ab} = 1$.

If a and b belong to uncertainty areas of two adjacent functions $[0 < \mu_l(a) < 1, 0 < \mu_{l+1}(a) < 1, 0 < \mu_l(b) < 1, 0 < \mu_{l+1}(b)]$, or one of these values belongs to tolerance area of one function, and another value – to uncertainty area of the adjacent function, then

$$\rho_{ab} = 1 - |\mu_l(a) - \mu_l(b)|. \tag{2.5}$$

The latter formula follows from characteristics of COSS term-set membership functions

$$\left. \begin{array}{l} \mu_l(a) + \mu_{l+1}(a) = 1; \\ \mu_l(b) + \mu_{l+1}(b) = 1 \Rightarrow \mu_l(a) - \mu_l(b) = \\ \mu_{l+1}(b) - \mu_{l+1}(a) \Rightarrow |\mu_l(a) - \mu_l(b)| = |\mu_{l+1}(a) - \mu_{l+1}(b)|. \end{array} \right\} \tag{2.6}$$

When comparing formulas (2.7) and (2.5), one can see that they are very similar, though the first formula operates with the parameter values, the second one — with values of their membership functions. The first formula includes a potency (length) of a characteristic value range equal to ($B - A$), the second one includes a potency of a range of membership functions values equal to 1. In other situations $\rho_{ab} = 0$.

Using COSS membership functions "inlet steam pressure" shown in Fig. 1.8., analogy degree of products with values of steam pressure 1.1 and 1.5, and analogy degree of products with values of steam pressure 6.1 and 6.6 were calculated.

Further results are obtained: $\rho_{1,1;1,5} = 0,34$, $\rho_{6,1;6,6} = 0,82$. According to [35], these results are consistent with the experience of experts.

Example 2.9. Fuzzy control of technological process of coal beneficiation by a separation method. The separation method has been widely spread in coal preparation industry throughout the world thanks to it simplicity, cheapness and universality. From the automation point of view, separation process is complicated enough due to absence of adequate mathematical model and presence of a number of difficult-to-control variations (load, granulometric, fractional and chemical composition of a raw product), and also lack of operative analysis methods of prepared product and wastes.

Hydraulic separation is a process of separation of raw mixture components in vertically pulsing stream of water having an alternating velocity. According to a classical trend of the separation theory, the separation coefficient during coal beneficiation in the jolting machine is the relative difference in fall final velocities depending particle size, density and medium density. Raw coal is loaded on a sieve of work bay. Piston or compressed air initiates vertical oscillations of water. A mixture of coal, breed and intermediate fractions, which locates on the sieve (bed), is entered alternately in loosened and condensed conditions by pulsing water stream. Due to press of portions of loaded raw material and also due to movement of transport water arriving to the machine with a raw material and under-sieve water, the whole bed moves in a horizontal direction being simultaneously stratified on heavy and light products. The lower layers of bed consisting of heavy products are removed from the jolting machine through a discharge unit which located on ends of each stage of the jolting machine. The prepared product or a concentrate together with water are discharged through a decanting edge.

The parameters influencing separation process are either hydrodynamic or technological. The hydrodynamic parameters are those initiating oscillation conditions of medium and fluidization of bed (flow rate of under-sieve water and air flow rate are of especially great importance). Technological parameters are connected with quality of prepared coal, i.e. their fractional and granulometric composition.

The coal beneficiation fuzzy control model for a separation method is developed by the laboratory of fuzzy techniques of Institute of an automation and electrometry of the Siberian branch of the Russian Academy of Sciences.

The control scheme proposed has two hierarchy levels. Lower level consists of technological process parameter regulators (unloading of heavy fractions, air and water supply). Operation mode of these regulators is set by a top-level fuzzy controller based on the information on structure of raw working mixture and operative control data (of fractional analysis, indications of ash gauge, etc.). The bed level regulator is implemented as two-input fuzzy controller. A bed height error signal (difference between set value and actual value) and a bed height error derivative are delivered to an input. Stimuli governing unload velocity are delivered to an output. Linguistic values of bed height error and its derivative are following: high negative (HN), low negative (LN), zero (Z), low positive (LP), high positive (HP). Linguistic values of unload velocity regulation are following: tremendously reduce (TR), strongly reduce (SR), moderately reduce (MR), slightly reduce (SLR), not to change (NC), slightly increase (SLI), moderately increase (MI), strongly increase (SI), tremendously increase (TI).

Table 2.2 Base of fuzzy logic conclusion rules

Error	Derivative of the error	Regulation of unload velocity
HN	HN	TR
HN	LN	SR
HN	Z	MR
HN	LP	SLR
HN	HP	NC
LN	HN	SR
LN	LN	MR
LN	Z	SLR
LN	LP	NC
LN	HP	SLI
Z	HN	MR
Z	LN	SLR
Z	Z	NC
Z	LP	SLI
Z	HP	MI
LP	HN	SLR
LP	LN	NC
LP	Z	SLI
LP	LP	MI
LP	HP	SI
HP	HN	NC
HP	LN	SLI
HP	Z	MI
HP	LP	SI
HP	HP	TI

Formalization of input and output linguistic values is carried out based on the COSS terms. Controlling output value is formed by fuzzy logic conclusion in accordance with the rule base provided in Table 2.2.

Efficiency of the model was proved experimentally under various technological process modes.

Chapter 3

Methods of Comparative and Fuzzy Cluster Analysis of Formalized Information

3.1 Definition of Comparative Indicators and Indicators of Models of Expert Characteristic Evaluations Consistency

As mentioned in Chapter 1, one cannot exclude that while estimating the same characteristic at a population of objects by several experts, the information obtained from the experts will differ various. As models of expert evaluations of some characteristic are constructed based on this information, they will obvious differ.

Thus, if for one characteristic several models can be constructed, then we have a task of their comparative analysis with the subsequent building of a generalized model.

Let us denote with Ξ^k a set the elements of which are k models of expert evaluations of qualitative characteristic or expert description of physical values of quantitative characteristic in linguistic terms (k COSS)

$$X_i = \left\{ \mu_{il}(x), l = \overline{1,m} \right\}, \ i = \overline{1,k} \ , \ \mu_{il}(x) \equiv \left(a_1^{il}, a_2^{il}, a_L^{il}, a_R^{il} \right)$$

where $\left[a_1^{il}, a_2^{il} \right]$ is a tolerance interval; a_L^{il}, a_R^{il} are the left and right parameters of a fuzziness, accordingly.

After determining of set Ξ^k it is necessary to carry out comparative analysis of its elements. In particular, it is necessary to find to what extent the elements are various (similar) pairwise, to what extent all elements are various (similar) in aggregate, whether there are essentially differing groups of elements, or structural composition of set Ξ^k is homogeneous enough.

For this purpose comparative quantity indicators of expert characteristic evaluation models are defined, based on which the fuzzy binary relations of similarity and conformity are then constructed.

As is known (15), a fuzzy set $\tilde{C}$ is referred to as intersection of fuzzy sets $\tilde{A}$ and $\tilde{B}$, $\tilde{C} = \tilde{A} \cap \tilde{B}$, if $\mu_{\tilde{C}}(x) = \mu_{\tilde{A}}(x) \wedge \mu_{\tilde{B}}(x)$, $\forall x \in X$ where $\wedge$ is an operator of triangular norm class.

The fuzzy set $\tilde{C}$ is referred to as association of fuzzy sets $\tilde{A}$ and $\tilde{B}$, $\tilde{C} = \tilde{A} \cup \tilde{B}$, if $\mu_{\tilde{C}}(x) = \mu_{\tilde{A}}(x) \vee \mu_{\tilde{B}}(x)$, $\forall x \in X$, where $\vee$ is an operator of triangular conorm class.

O.M. Poleshchuk and E.G. Komarov: Expert Fuzzy Info. Processing, STUDFUZZ 268, pp. 87–117.
springerlink.com   © Springer-Verlag Berlin Heidelberg 2011

Based on these known definitions, let us define operations for set Ξ^k elements, which form semantic space with term-sets membership functions.

With intersection of two elements $X_i \cap X_j$; $i = \overline{1,k}$; $j = \overline{1,k}$

$$\left\{\mu_l(x) = \min\left[\mu_{il}(x), \mu_{jl}(x)\right]; \ \forall x; \ l = \overline{1,m}\right\}$$

With intersection of elements $X_1 \cap ... \cap X_k$

$$\left\{\min_{i=1,k}\left[\mu_{1l}(x),...,\mu_{kl}(x)\right]; \ \forall x\right\}$$

With union of two elements $X_i \cup X_j$

$$\left\{\mu_l(x) = \max_{i=1,k}\left[\mu_{il}(x), \mu_{jl}(x)\right]; \ \forall x\right\}.$$

With union of elements $X_1 \cup ... \cup X_k$

$$\left\{\mu_l(x) = \max_{i=1,k}\left[\mu_{1l}(x),...,\mu_{kl}(x)\right]; \ \forall x\right\}.$$

With generalized sum of two elements $X_i + X_j$

$$\left\{\mu_l(x) \equiv \left(a_1^{il} + a_1^{jl}, a_2^{il} + a_2^{jl}, a_L^{il} + a_L^{jl}, a_R^{il} + a_R^{jl}\right)\right\}$$

With generalized product of an element X_i with a positive number C for COSS

$$\left\{\mu_l(x) \equiv \left(ca_1^{il}, ca_2^{il}, ca_L^{il}, ca_R^{il}\right)\right\}$$

With generalized linear combination $\sum_{i=1}^{k} c_i X_i$, $c_i > 0$ for COSS

$$\left\{\mu_l(x) \equiv \left(\sum_{i=1}^{k} c_i a_1^{il}, \sum_{i=1}^{k} c_i a_2^{il}, \sum_{i=1}^{k} c_i a_L^{il}, \sum_{i=1}^{k} c_i a_R^{il}\right)\right\}.$$

Let us assume that semantic space Y with membership functions of term-set $\eta_l(x)$ belongs to semantic space Z with membership functions of term-set $v_l(x)$, if the following conditions are satisfied

$$\eta_l(x) \le v_l(x), \ \forall l = \overline{1,m}.$$

Let us define the quantity indexes characterizing distinctions or similarities of two elements of set Ξ^k with membership functions $\{\mu_{il}(x), \ l = \overline{1,m}\}$; $\{\mu_{jl}(x)\}$ for universal set $U = [0,1]$. If universal set is a segment $U = [a,b]$, then it is necessary to primarily reduce parameters of membership functions of elements to [0.1] by the formula

$$\mu_{il}(x) \equiv \left(\frac{a_1^{il} - a}{b - a}; \ \frac{a_2^{il} - a}{b - a}; \ \frac{a_L^{il}}{b - a}; \ \frac{a_R^{il}}{b - a}\right).$$

Let us define index of distinction within the limits of l-th term of two elements as loss of the information within the scope of l-th term between these elements

$$d\left(\mu_{il}, \mu_{jl}\right) = \int_0^1 \left|\mu_{il}(x) - \mu_{jl}(x)\right| dx.$$

Index of similarity within the scope of l-th term

$$\tilde{\kappa}_{ij}^l = 1 - d\left(\mu_{il}, \mu_{jl}\right).$$

Let us define index of distinction of two elements as information loss between these elements

$$d\left(X_i, X_j\right) = \frac{1}{2} \sum_{l=1}^m \int_0^1 \left|\mu_{il}(x) - \mu_{jl}(x)\right| dx. \tag{3.1}$$

Index of similarity of two elements

$$\tilde{\kappa}_{ij} = 1 - d\left(X_i, X_j\right). \tag{3.2}$$

Index of consistency of two elements within the scope of l-th term

$$\kappa_{ij}^l = \frac{\int_0^1 \min\left[\mu_{il}(x), \mu_{jl}(x)\right] dx}{\int_0^1 \max\left[\mu_{il}(x), \mu_{jl}(x)\right] dx}.$$

Index of consistency of two elements

$$\kappa_{ij} = \frac{1}{m} \sum_{l=1}^m \frac{\int_0^1 \min\left[\mu_{il}(x), \mu_{jl}(x)\right] dx}{\int_0^1 \max\left[\mu_{il}(x), \mu_{jl}(x)\right] dx}.$$

All indexes vary from zero to unit.

While solving practical problems the selection of indexes of similarity or consistency depends on a task set.

If it is necessary to determine degree of similarity between two elements of set Ξ^k (of two COSS's) within the scope of terms with the greatest carriers, a similarity index is recommended to use; and if within the scope of all terms irrespective of their carriers, a consistency index is recommended to use.

If COSS's are constructed based on direct inquiries of i-th and j-th experts by the methods stated in §2.4, 2.5, then indexes defined above are treated accordingly as distinction (similarity, consistency) indexes of individual criteria of i-th and j-th experts or description of characteristics.

If COSS's are constructed to the methods stated in §2.2 and 2.3 based on the evaluations of a population of objects within the scope of some qualitative characteristic given by i-th and j-th experts, the indexes defined above are treated accordingly as distinction (similarity, consistency) indexes of criteria of i-th and j-th experts.

If COSS's are constructed to the methods stated in § 2.2 and 2.3 based on the expert evaluation of appearances of qualitative characteristics of i-th and j-th populations of objects, the indexes defined above are treated accordingly as distinction (similarity, consistency) indexes of appearance of this particular qualitative characteristic.

Let us identify indexes of the general consistency of elements of set Ξ^k with membership functions, accordingly

$$\kappa = \frac{1}{m}\sum_{l=1}^{m} \frac{\int\limits_0^1 \min[\mu_{1l}(x),\mu_{2l}(x),...,\mu_{kl}(x)]dx}{\int\limits_0^1 \max[\mu_{1l}(x),\mu_{2l}(x),...,\mu_{kl}(x)]dx}; \qquad (3.3)$$

$$\tilde{\kappa} = \sqrt[m]{\prod_{l=1}^{m} \frac{\int\limits_0^1 \min[\mu_{1l}(x),\mu_{2l}(x),...,\mu_{kl}(x)]dx}{\int\limits_0^1 \max[\mu_{1l}(x),\mu_{2l}(x),...,\mu_{kl}(x)]dx}}. \qquad (3.4)$$

The index κ is referred to as an additive index of consistency of elements of set Ξ^k. The index $\tilde{\kappa}$ is referred to as a multiplicative index of consistency of elements of set Ξ^k. The variation range of both indexes is a segment [0, 1]. If all elements of set Ξ^k coincide, indexes are equal to unity. The additive index is equal to zero, when membership functions of all terms have no crosscuts; and multiplicative index is equal to zero, when membership functions of at least one term have no crosscuts.

3.2   The Fuzzy Cluster Analysis of Set of Models of Expert Characteristic Evaluations

Building of fuzzy binary relations of similarity and conformity on set of models of expert characteristic evaluations allows carrying out the fuzzy cluster analysis of this set and, by that, to study its structural composition.

The Proposition 3.1. [135] Fuzzy sets $R_1 - R_4$ with membership functions, accordingly

$$\mu_{R_1}(X_i, X_j) = \tilde{\kappa}_{ij}, \; \mu_{R_2}(X_i, X_j) = \kappa_{ij}, \; \mu_{R_3}(X_i, X_j) = \tilde{\kappa}_{ij}^l, \; \mu_{R_4}(X_i, X_j) = \kappa_{ij}^l,$$

where $i = \overline{1,k}$; $j = \overline{1,k}$; $l = \overline{1,m}$.

Define fuzzy binary relations of similarity over set Ξ^k .

The proof. Let us prove of reflexivity characteristic for $R_p \left(p = \overline{1,4}\right)$, i.e. that

$$\mu_{R_p}\left(X_i, X_i\right) = 1_{;} \; X_i \in \Xi^k$$

$$\mu_{R_1}\left(X_i, X_i\right) = \tilde{\kappa}_{ii} = 1 - d\left(X_i, X_i\right) = 1 - \frac{1}{2}\sum_{l=1}^{m}\int_0^1\left|\mu_{il}(x) - \mu_{il}(x)\right|dx = 1;$$

$$\mu_{R_2}\left(X_i, X_i\right) = \kappa_{ii} = \frac{1}{m}\sum_{l=1}^{m}\frac{\displaystyle\int_0^1\min\left[\mu_{il}(x), \mu_{il}(x)\right]dx}{\displaystyle\int_0^1\max\left[\mu_{il}(x), \mu_{il}(x)\right]dx} = 1;$$

$$\mu_{R_3}\left(X_i, X_i\right) = \tilde{\kappa}_{ii}^l = 1 - d\left(\mu_{il}, \mu_{il}\right) = 1 - \int_0^1\left|\mu_{il}(x) - \mu_{il}(x)\right|dx = 1;$$

$$\mu_{R_4}\left(X_i, X_i\right) = \kappa_{ii}^l = \frac{\displaystyle\int_0^1\min\left[\mu_{il}(x), \mu_{il}(x)\right]dx}{\displaystyle\int_0^1\max\left[\mu_{il}(x), \mu_{il}(x)\right]dx} = 1.$$

Thus, R_p is a set with reflective fuzzy relations. Let us prove symmetry characteristic for this set, i.e. that $\mu_{R_p}\left(X_i, X_j\right) = \mu_{R_p}\left(X_j, X_i\right)$; $X_i, X_j \in \Xi^k$

$$\mu_{R_1}\left(X_i, X_j\right) = 1 - d\left(X_i, X_j\right) = 1 - \frac{1}{2}\sum_{l=1}^{m}\int_0^1\left|\mu_{il}(x) - \mu_{jl}(x)\right|dx =$$

$$= 1 - \frac{1}{2}\sum_{l=1}^{m}\int_0^1\left|\mu_{jl}(x) - \mu_{il}(x)\right|dx = 1 - d\left(X_j, X_i\right) = \mu_{R_1}\left(X_j, X_i\right);$$

$$\mu_{R_2}\left(X_i, X_j\right) = \frac{1}{m}\sum_{l=1}^{m}\frac{\displaystyle\int_0^1\min\left[\mu_{il}(x), \mu_{jl}(x)\right]dx}{\displaystyle\int_0^1\max\left[\mu_{il}(x), \mu_{jl}(x)\right]dx} =$$

$$= \frac{1}{m}\sum_{l=1}^{m}\frac{\displaystyle\int_0^1\min\left[\mu_{jl}(x), \mu_{il}(x)\right]dx}{\displaystyle\int_0^1\max\left[\mu_{jl}(x), \mu_{il}(x)\right]dx} = \mu_{R_2}\left(X_j, X_i\right);$$

$$\mu_{R_3}\left(X_i, X_j\right) = 1 - d\left(\mu_{il}, \mu_{jl}\right) = 1 - \int_0^1\left|\mu_{il}(x) - \mu_{jl}(x)\right|dx =$$

$$= 1 - \int_0^1 \left| \mu_{jl}(x) - \mu_{il}(x) \right| dx = 1 - d\left(\mu_{jl}, \mu_{il} \right) = \mu_{R_3}\left(X_j, X_i \right);$$

$$\mu_{R_4}\left(X_i, X_j \right) = \frac{\int_0^1 \min\left[\mu_{il}(x), \mu_{jl}(x) \right] dx}{\int_0^1 \max\left[\mu_{il}(x), \mu_{jl}(x) \right] dx} =$$

$$= \frac{\int_0^1 \min\left[\mu_{jl}(x), \mu_{il}(x) \right] dx}{\int_0^1 \max\left[\mu_{jl}(x), \mu_{il}(x) \right] dx} = \mu_{R_4}\left(X_j, X_i \right).$$

Thus, R_p are symmetric fuzzy relations. From characteristics of reflexivity and symmetry it follows that R_p are fuzzy relations of similarity. The proposition 3.1 is proved.

Generally, constructed fuzzy relations of similarity R_p are not transitive. Let us denote fuzzy relations of similarity which are transitive closures of fuzzy relations R_p with $\hat{R}_p$.

Let us identify elements $X_1, X_2, ..., X_q$, $q \le k$ of set Ξ^k as conform (concerning the conformity relation $\hat{R}_p$) with confidence level $\alpha \in (0,1)$, if for all X_i, X_j at $i = \overline{1,q}$; $j = \overline{1,q}$ the relation $\mu_{\hat{R}}\left(X_i, X_j \right) \ge \alpha$ is satisfied.

Let us consider fuzzy clusterization of a set Ξ^k under relations of conformity $\hat{R}_p$.

For relations

$$\mu_{R_1}\left(X_i, X_j \right) = \tilde{\kappa}_{ij}, \ \mu_{R_2}\left(X_i, X_j \right) = \tilde{\kappa}_{ij}^1, \ \mu_{R_3}\left(X_i, X_j \right) = \kappa_{ij}, \ \mu_{R_4}\left(X_i, X_j \right) = \kappa_{ij}^1$$

Let us write out corresponding matrixes of fuzzy relations of similarity: R

$$\mathrm{R}_p = \begin{pmatrix} 1 & \mu_{R_p}\left(X_1, X_2 \right) & \cdots & \mu_{R_p}\left(X_1, X_k \right) \\ \mu_{R_p}\left(X_1, X_2 \right) & 1 & \cdots & \mu_{R_p}\left(X_2, X_k \right) \\ \cdot & \cdot & \cdots & \cdot \\ \mu_{R_p}\left(X_1, X_k \right) & \mu_{R_p}\left(X_2, X_k \right) & \cdots & 1 \end{pmatrix}. \tag{3.5}$$

As generally, constructed fuzzy relations of similarity with matrixes (3.6) are not transitive, let us construct transitive closures of similarity relations R_p based on the union of compositions of each considered relations with themselves. The possibility of such building methods is considered in [15]. Moreover, since R_p are reflective fuzzy binary relations (the proposition 3.1), from [15] it follows that $\hat{R}_p = R_p^{k-1}$.

Let $\hat{R}_p$ be fuzzy relations of similarity which are transitive closures of fuzzy relations of similarity R_p, they are defined by matrixes

$$\hat{R}_p = \begin{pmatrix} 1 & \mu_{\hat{R}_p}(X_1, X_2) & \cdots & \mu_{\hat{R}_p}(X_1, X_k) \\ \mu_{\hat{R}_p}(X_1, X_2) & 1 & \cdots & \mu_{\hat{R}_p}(X_2, X_k) \\ \cdot & \cdot & \cdots & \cdot \\ \mu_{\hat{R}_p}(X_1, X_k) & \mu_{\hat{R}_p}(X_2, X_k) & \cdots & 1 \end{pmatrix}.$$

Thus, $\hat{R}_p$ define within Ξ^k the fuzzy binary relations of similarity, and, accordingly, hierarchy of partitioning of set of expert indicator evaluation models by equivalence classes.

That is, according to the decomposition theorem for similarity relations $\hat{R}_p$ it is possible to decompose:

$$\hat{R}_p = \bigcup_\alpha \left\{ \alpha \begin{pmatrix} 1 & \delta_{12} & \cdots & \delta_{1k} \\ \delta_{12} & 1 & \cdots & \delta_{2k} \\ \cdot & \cdot & \cdots & \cdot \\ \delta_{1k} & \delta_{2k} & \cdots & 1 \end{pmatrix} \right\}, \tag{3.6}$$

onto equivalence relations, where $\delta_{ij} = \begin{cases} 0; \\ 1, \end{cases} i = \overline{1, k}, \ j = \overline{1, k}.$

Populations of units form matrix partitions on minors corresponding to clusters.

Thus, depending on α-levels of fuzzy relations of similarity, the set Ξ^k can be divided into clusters of elements similar among themselves with α levels of confidence.

An application of the method of fuzzy cluster analysis of set Ξ^k is check and rejection of erroneous information. Under conditions of unsatisfactory consistency indexes of elements of this set (value 0.5 can be a threshold of satisfactory consistency), it is possible to select falling out elements (models of expert evaluations of qualitative characteristics or expert description in linguistic terms of quantitative characteristics). The invalidity of information of these models is confirmed by raised concordance indexes of elements of set Ξ^k when being rejected.

3.3 Building of Comparative Indexes and Indexes of Consistency of Formalized Outcomes of an Qualitative Characteristic Evaluation for a Population of Objects

Let us assume that k experts estimate appearance of qualitative characteristic at a population of objects. According to the methods stated in §2.2 and 2.3, k COSS X_i; $i = \overline{1, k}$ can be constructed with membership functions of term-sets

$\{\mu_{il}(x); l = \overline{1,m}\}$. These COSS's are elements of set Ξ^k. Let us consider that for each object the expert evaluations are known, and they are levels (terms) of a verbal scale over which the evaluation is performed. Thus, if a point given by i-th expert to n-th object $\left(n = \overline{1,N}\right)$ is l-th level of a verbal scale, it is unambiguously mapped on membership function $\mu_{il}(x)$ of l-th COSS term. Let us denote a point given by i-th expert to n-th representative of a population $\mu_i^n(x) = \mu_{il}(x)$, and a collection of the formalized evaluations given by i-th expert

$$M_i = \{\mu_i^1(x), \mu_i^2(x), ..., \mu_i^N(x)\}; \ \mu_i^n(x) \equiv \left(a_1^{in}, a_2^{in}, a_L^{in}, a_R^{in}\right),$$

where $\mu_i^n(x)$ is one of membership functions of i-th element $X_i = \mu_{il}(x)$ of set Ξ^k.

Let us introduce some legends for operations with elements M_i of set Θ^k, for which a collection N of fuzzy numbers with membership functions is obtained: with cross-section of two elements $M_i \cap M_j$

$$\{\mu^n(x)\} = \{\min[\mu_i^n(x), \mu_j^n(x)], \forall x\};$$

with cross-section of elements $M_1 \cap ... \cap M_k$

$$\{\mu^n(x)\} = \left\{\min_{i=1,k}[\mu_1^n(x), ..., \mu_k^n(x)], \forall x\right\}$$

with union of two elements $M_i \cup M_j$

$$\{\mu^n(x)\} = \{\max[\mu_i^n(x), \mu_j^n(x)], \forall x\};$$

with union of elements $M_1 \cup ... \cup M_k$

$$\{\mu^n(x)\} = \left\{\max_{i=1,k}[\mu_1^n(x), ..., \mu_k^n(x)], \forall x\right\};$$

For generalized sum of two elements $M_i + M_j$

$$\{\mu_i^n(x) \equiv \left(a_1^{in} + a_1^{jn}, a_2^{in} + a_2^{jn}, a_L^{in} + a_L^{jn}, a_R^{in} + a_R^{jn}\right)\};$$

For generalized product of an element M_i and positive value c

$$\{\mu_i^n(x) \equiv \left(ca_1^{in}, ca_2^{in}, ca_L^{in}, ca_R^{in}\right)\};$$

For generalized linear combination $\sum\limits_{i=1}^{k} c_i M_i \ (c_i > 0)$

$$\left\{ \mu^n(x) \equiv \left(\sum_{i=1}^{k} c_i a_1^{in}, \sum_{i=1}^{k} c_i a_2^{in}, \sum_{i=1}^{k} c_i a_L^{in}, \sum_{i=1}^{k} c_i a_R^{in} \right) \right\}.$$

On constructing the set Θ^k, the elements of which are the formalized results of M_i expert evaluations of qualitative characteristic at a population of objects, it is necessary to be capable to carrying out the comparative analysis of its elements: to what extent those elements are pairwise various (similar); to what extent all elements are various (similar) in aggregate; whether there are essentially differing groups of elements, or structural composition of set Θ^k is homogeneous enough.

Let us define quantity indexes for universal set $U = [0,1]$. If the universal set is a segment $U = [a,b]$, then, it is necessary to preliminary reduce to [0, 1] the parameters of membership functions of set elements based on which the elements of set Θ^k are constructed, by the formula:

$$\mu_{il}(x) \equiv \left(\frac{a_1^{il} - a}{b-a}, \frac{a_2^{il} - a}{b-a}, \frac{a_L^{il}}{b-a}, \frac{a_R^{il}}{b-a} \right); \ l = \overline{1,m}.$$

To determine consistency of two and more expert evaluations, various indexes are applied; they are based either on paired comparisons of the rating points assigned to the same objects (for example, Kendall coefficient [137]), or on arithmetical transformations of rating points (for example, concordance coefficients [138], a grade correlation in Kemeni-Snell model [139], Spearman grade correlation [140]). Other indexes based on the principles similar to those above are stated in [141—145].

All defined below quantity indexes between i-th and j-th elements of set Θ^k are defined on the basis of abstract concepts, namely, values of membership functions [135], and operations between membership functions are defined on the basis of minimax operators $(i = 1, k; \ j = 1, k)$

Distinction index

$$d(M_i, M_j) = \frac{1}{N} \sum_{n=1}^{N} \int_0^1 |\mu_i^n(x) - \mu_j^n(x)| dx;$$

Similarity index

$$\hat{k}_{ij} = 1 - d(M_i, M_j);$$

(3.7)

Consistency index

$$k_{ij} = \frac{1}{N} \sum_{n=1}^{N} \frac{\int_0^1 \min[\mu_i^n(x), \mu_j^n(x)] dx}{\int_0^1 \max[\mu_i^n(x), \mu_j^n(x)] dx};$$

(3.8)

Additive and multiplicative indexes, accordingly

$$\hat{k}_{ij} = \frac{1}{N}\sum_{n=1}^{N} \frac{\int_0^1 \min[\mu_1^n(x),\mu_2^n(x),...,\mu_k^n(x)]dx}{\int_0^1 \max[\mu_1^n(x),\mu_2^n(x),...,\mu_k^n(x)]dx};$$

(3.9)

$$\tilde{k} = \sqrt[N]{\prod_{n=1}^{N} \frac{\int_0^1 \min_x(\mu_1^n,\mu_2^n,...,\mu_k^n)dx}{\int_0^1 \max_x(\mu_1^n,\mu_2^n,...,\mu_k^n)}}.$$

(3.10)

All indexes vary from zero to unity. If values of consistency indexes are close to zero, it means incompetence of at least several experts or it means a fuzzy proposition of evaluation procedure.

3.4 The Fuzzy Cluster Analysis of Set of the Formalized Results of an Evaluation of Qualitative Characteristic of a Population of Objects

Building of fuzzy binary relations of similarity and conformity over a set of formalized results of expert evaluations allows to carry out fuzzy cluster analysis of this set and, by that, to study its structural composition.

The proposition 3.2. [135] Fuzzy sets R_1, R_2 with membership functions $\mu_{R_1}(M_i, M_j) = k_{ij}$, $\mu_{R_2}(M_i, M_j) = \tilde{k}_{ij}$ $(i = \overline{1,k}; \; j = \overline{1,k})$, accordingly, define fuzzy relations of similarity over set Θ^k.

The proof. Let us prove that R_1 and R_2 possess characteristics of reflexivity and symmetry

$$\mu_{R_1}(M_i, M_i) = \tilde{k}_{ii} = 1 - d(M_i, M_i)\Bigl[1 - \frac{1}{N}\sum_{n=1}^{N}\int_0^1 |\mu_i^n(x) - \mu_i^n(x)|dx = 1;$$

$$\mu_{R_2}(M_i, M_i) = k_{ij} = \frac{1}{N}\sum_{n=1}^{N} \frac{\int_0^1 \min[\mu_i^n(x),\mu_i^n(x)]dx}{\int_0^1 \max[\mu_i^n(x),\mu_i^n(x)]dx} = 1.$$

Thus, R_1 and R_2 are reflective. Let us prove their symmetry

$$\mu_{R_1}(M_i, M_j) = 1 - d(M_i, M_j) = 1 - \frac{1}{N}\sum_{n=1}^{N}\int_0^1 |\mu_i^n(x) - \mu_j^n(x)|dx =$$

$$= 1 - \frac{1}{N} \sum_{n=1}^{N} \int_0^1 \left| \mu_j^n(x) - \mu_i^n(x) \right| dx = \mu_{R_1}\left(M_j, M_i\right);$$

$$\mu_{R_2}\left(M_i, M_j\right) = \frac{1}{N} \sum_{n=1}^{N} \frac{\int_0^1 \min\left[\mu_i^n(x), \mu_j^n(x)\right] dx}{\int_0^1 \max\left[\mu_i^n(x), \mu_j^n(x)\right] dx} =$$

$$= \frac{1}{N} \sum_{n=1}^{N} \frac{\int_0^1 \min\left[\mu_j^n(x), \mu_i^n(x)\right] dx}{\int_0^1 \max\left[\mu_j^n(x), \mu_i^n(x)\right] dx} = \mu_{R_2}\left(M_j, M_i\right).$$

Thus, R_1 and R_2 are symmetric. The proposition 3.2 is proved.

If the constructed fuzzy relations of similarity R_1, R_2 are not transitive, let us denote by $\hat{R}_1, \hat{R}_2$, accordingly, fuzzy relations of similarity which are transitive closures of fuzzy relations R_1, R_2. Since in accordance with the proposition 3.2 R_1, R_2 are reflective fuzzy relations, then $\hat{R}_p = R_p^{k-1}$, $p = \overline{1,2}$.

Let us denote elements $M_1, M_2, ..., M_q$ $(q \leq k)$ of the set Θ^k as similar (versus similarity relation $\hat{R}$) with confidence level $\alpha \in (0,1)$, if for all M_i, M_j $\left(i = \overline{1,q}; j = \overline{1,q}\right)$ the relation $\mu_{\hat{R}}\left(M_i, M_j\right) \geq \alpha$ is satisfied.

Let us consider fuzzy clusterization of a set Θ^k under fuzzy relations of similarity $\hat{R}_1, \hat{R}_2$.

In the proposition 3.2 it is proved that $\mu_{R_1}\left(M_i, M_j\right) = \tilde{k}_{ij}$, $\mu_{R_2}\left(M_i, M_j\right) = \tilde{k}_{ij}$ $\left(i = \overline{1,k}; j = \overline{1,k}\right)$ are values of membership functions of fuzzy relations of similarity R_1, R_2 defined over set Θ^k. Let us make a matrix of fuzzy relations of similarity for these relations:

$$R_p = \begin{pmatrix} 1 & \mu_{R_p}(M_1, M_2) & \cdots & \mu_{R_p}(M_1, M_k) \\ \mu_{R_p}(M_1, M_2) & 1 & \cdots & \mu_{R_p}(M_2, M_k) \\ \cdot & \cdot & \cdots & \cdot \\ \mu_{R_p}(M_1, M_k) & \mu_{R_p}2(M_2, M_k) & \cdots & 1 \end{pmatrix}, \; p = \overline{1,2}.$$

Generally, constructed fuzzy relations of similarity R_1, R_2 are not transitive, therefore let us construct their transitive closures $\hat{R}_1, \hat{R}_2$, which are fuzzy relations of similarity. Transitive closures for R_1, R_2 are constructed on the basis of compositions of each considered relation with themselves. Let us write out matrixes of fuzzy relations of similarity

$$\hat{R}_p = \begin{pmatrix} 1 & \mu_{\hat{R}_p}(M_1, M_2) & \cdots & \mu_{\hat{R}_p}(M_1, M_k) \\ \mu_{\hat{R}_p}(M_1, M_2) & 1 & \cdots & \mu_{\hat{R}_p}(M_2, M_k) \\ \cdot & \cdot & \cdots & \cdot \\ \mu_{\hat{R}_p}(M_1, M_k) & \mu_{\hat{R}_p} 2(M_2, M_k) & \cdots & 1 \end{pmatrix}, \quad p = \overline{1,2}, \tag{3.11}$$

which define over Θ^k fuzzy binary relations of similarity and, accordingly, hierarchy of partitioning of set of the formalized results of an qualitative index expert evaluation for a population of objects by equivalence classes.

According to (3.12), it is possible to decompose matrixes of relations of similarity $\hat{R}_p$ onto equivalence relations:

$$\hat{R}_p = \bigcup_\alpha \left\{ \alpha \begin{pmatrix} 1 & \delta_{12} & \cdots & \delta_{1k} \\ \delta_{12} & 1 & \cdots & \delta_{2k} \\ \cdot & \cdot & \cdots & \cdot \\ \delta_{1k} & \delta_{2k} & \cdots & 1 \end{pmatrix} \right\}, \tag{3.12}$$

where $\delta_{ij} = \begin{cases} 0; \\ 1. \end{cases}$

Populations of objects form partitions of the matrix onto minors corresponding to clusters.

Thus, depending on α-levels of fuzzy similarity relations, the set Θ^k can be divided into clusters of similar elements with levels of confidence α.

3.5   Examples of Practical Application of the Fuzzy Cluster Analysis Methods

Example 3.1. [146] **Assignment of boards of examiners.** Quality of checks of entrants' exam papers directly depends on to what extent accurate and coordinated knowledge evaluation criteria of the examiners appointed to the subject boards are. To solve a problem of assigning the boards of examiners the method based on the fuzzy cluster analysis is offered. Let us consider pupils' knowledge on mathematics estimated results provided by five examiners within the limits of a scale "E","C", "B", "A", summarized in Table 3.1

Table 3.1 Estimated results of knowledge on mathematics provided by examiners

No. of an examiner	A ("5")	B ("4")	C ("3")	E ("2")
1	48	110	90	36
2	50	87	120	27
3	50	99	92	43
4	48	100	97	39
5	42	99	93	50

Based on method described in §2.2 and data of Table 3.1, membership functions of the formalized individual criteria of examiners μ_{il} $\left(i = \overline{1,5};\ l = \overline{1,4}\right)$ are obtained, having the following parameters:

$$\mu_{11} = (0;\, 0,063;\, 0;\, 0,126).\quad \mu_{12} = (0,189;\, 0,285;\, 0,126;\, 0,318).$$

$$\mu_{13} = (0,603;\, 0,7465;\, 0,318;\, 0,169).\quad \mu_{14} = (0,9155;\, 1;\, 0,169;\, 0).$$

$$\mu_{21} = (0;\, 0,048;\, 0;\, 0,096).\quad \mu_{22} = (0,144;\, 0,365;\, 0,096;\, 0,306).$$

$$\mu_{23} = (0,671;\, 0,736;\, 0,306;\, 0,176).\quad \mu_{24} = (0,912;\, 1;\, 0,176;\, 0).$$

$$\mu_{31} = (0;\, 0,0765;\, 0;\, 0,153).\quad \mu_{32} = (0,2295;\, 0,315;\, 0,153;\, 0,324).$$

$$\mu_{33} = (0,639;\, 0,7375;\, 0,324;\, 0,175).\quad \mu_{34} = (0,9125;\, 1;\, 0,175;\, 0).$$

$$\mu_{41} = (0;\, 0,0685;\, 0;\, 0,137).\quad \mu_{42} = (0,2055;\, 0,3075;\, 0,137;\, 0,341).$$

$$\mu_{43} = (0,6485;\, 0,748;\, 0,341;\, 0,168).\quad \mu_{44} = (0,916;\, 1;\, 0,168;\, 0).$$

$$\mu_{51} = (0;\, 0,086;\, 0;\, 0,172).\quad \mu_{52} = (0,258;\, 0,3355;\, 0,172;\, 0,327).$$

$$\mu_{53} = (0,6625;\, 0,7735;\, 0,327;\, 0,151).\quad \mu_{54} = (0,9245;\, 1;\, 0,151;\, 0)$$

Since the additive index of the general consistency of criteria (3.4) is equal to 0.705, it is possible to draw a conclusion that all examiners considered are competent.

Let us calculate indexes of pairwise similarity (3.2) of examiners' criteria. Results of evaluations are summarized in Table 3.2.

Table 3.2 Elements of a matrix of pairwise similarity of examiners' criteria

1	0,890	0,935	0,954	0,882
0,890	1	0,901	0,912	0,882
0,935	0,901	1	0,971	0,938
0,954	0,912	0,971	1	0,926
0,882	0,882	0,938	0,926	1

Based on the computed indexes of pairwise similarity of examiners' criteria, let us construct a fuzzy binary relation of similarity. Elements of the relation matrix are provided in Table 3.3.

Table 3.3 Elements of a matrix of fuzzy binary relation of examiners' criteria similarity

1	0,912	0,954	0,954	0,938
0,912	1	0,912	0,912	0,912
0,954	0,912	1	0,971	0,938
0,954	0,912	0,971	1	0,938
0,938	0,912	0,938	0,938	1

Using the decomposition theorem, the constructed relation of similarity is decomposed by equivalence relations, and a set of criteria is divided into clusters of criteria similar to each other with different confidence levels. The obtained results are given below.

Clusters of similar criteria of examiners with different levels

Confidence level	Cluster
0,912	{1,2,3,4,5}
0,938	{1,3,4,5},{2}
0,954	{1,3,4},{2},{5}
0,971	{3,4},{1},{2},{5}
1	{1},{2},{3},{4},{5}

Thus, if the board consists of five persons, then with rather high confidence level to the researches completed, all examiners can be appointed to the board. If the board consists of four persons it is offered to appoint examiners No. 1. 3. 4. 5. If the board consists of three or two persons, it is offered to appoint examiners No. 1, 3, and 4 or No. 3, 4, accordingly.

Example 3.2. [147] **Analysis of expert evaluation results of educational process quality.** As the input information, results of inquiry of experts within the limits of the following ten issues related to the correspondence of training quality to modern requirements are considered:

1. Courseware issued in a high school;
2. Teaching staff of a high school;
3. Teaching of students and availability of PC's in classrooms of a high school;
4. Curricula of special academic subjects;
5. Research activity of faculties and new tramlines of scientific thought in a high school;
6. Research activity of students covered by science and education integration approach;
7. Educational process in the field of future professional occupation of students;
8. Existing scale of an evaluation of students' knowledge, which influences quality of educational process;
9. Existing measures of students' knowledge evaluation control;
10. Forms of teaching.

Evaluations of experts are selected from three propositions: "mismatching", "partially corresponding", "corresponding" (Table 3.4).

Table 3.4 Relative results of expert opinions

Relative number	1	2	3	4	5	6	7	8	9	10
a_1^i	0,3	0,4	0,54	0,24	0,4	0,28	0,26	0,28	0,32	0,44
a_2^i	0,12	0,14	0,16	0,16	0,2	0,14	0,12	0,14	0,12	0,08
a_1^i	0,58	0,46	0,3	0,6	0,4	0,58	0,62	0,58	0,56	0,48

Note: a_1^i — "mismatching", a_2^i — "partially corresponding", a_1^i — "corresponding".

According to the obtained data a_l^i $\left(i = \overline{1,10}; l = \overline{1,3}\right)$ ten linguistic variables X_i = «attitude to i-th question» were constructed with membership functions denoted through $\mu_{il}(x)$, accordingly. Table 3.5 summarizes parameters of these functions.

Table 3.5 Parameters of membership functions of linguistic variables

Question	$\mu_{i1}(x)$	$\mu_{i2}(x)$	$\mu_{i3}(x)$
1	(0; 0,24; 0; 0,12)	(0,36; 0,12; 0,12)	(0,48; 1; 0,12; 0)
2	(0; 0,33; 0; 0,14)	(0,47; 0,14; 0,14)	(0,61; 1; 0,14; 0)
3	(0; 0,46; 0; 0,16)	(0,62; 0,16; 0,16)	(0,78; 1; 0,16; 0)
4	(0; 0,16; 0; 0,16)	(0,32; 0,16; 0,16)	(0,48; 1; 0,16; 0)
5	(0; 0,3; 0; 0,2)	(0,5; 0,2; 0,2)	(0,7; 1; 0,2; 0)
6	(0; 0,21; 0; 0,14)	(0,35; 0,14; 0,14)	(0,49; 1; 0,14; 0)
7	(0; 0,2; 0; 0,12)	(0,32; 0,12; 0,12)	(0,44; 1; 0,12; 0)
8	(0; 0,21; 0; 0,14)	(0,35; 0,14; 0,14)	(0,49; 1; 0,14; 0)
9	(0; 0,26; 0; 0,12)	(0,38; 0,12; 0,12)	(0,5; 1; 0,12; 0,12)
10	(0; 0,4; 0; 0,08)	(0,48; 0,08; 0,08)	(0,56; 1; 0,08; 0)

General consistency indexes for experts' opinions within the limits of all ten positions are the following: an additive one by (3.4) is equal to 309, a multiplicative one by (3.5) — 0.289. Values of these indexes mean that according to experts there are essential distinctions between evaluations of correspondence of the educational process quality in high school and modern training requirements within the scope of considered positions. Results of fuzzy clusterization of experts' opinions are shown below.

Confidence level	Cluster
0,59	{1,2,3,4,5,6,7,8,9,10}
0,63	{1,2,4,5,6,7,8,9,10}, {3}
0,81	{1,4,6,7,8,9}, {3}, {2,5,10}
0,83	{1,4,6,7,8,9}, {3}, {2,5}, {10}
0,88	{1,4,6,8,9}, {3}, {2}, {5}, {10}
0,92	{1,6,8}, {3}, {2}, {5}, {10},{4,7}, {9}
1	{1}, {2}, {3}, {4}, {5}, {6}, {7}, {8}, {9}, {10}

Thus, one can understand that the state of correspondence of computers (needed to ensure students' studies) availability in the classrooms of high school does not meet the appropriate requirement to the complete extent. State of correspondence of courseware issued by high school, research activity of students, integration of science and education spheres, and the existing scale of evaluation of students' knowledge to the current requirements of the training quality is more or less in line.

Example 3.3. The analysis of results of an expert evaluation of courseware quality. Six independent experts estimated characteristic «correspondences to curricula» of 20 samples of tutorials. The evaluation was performed within the limits of a verbal scale with six levels: "lack of correspondence", "correspondence in accessorial sections", "correspondence in basic sections", "correspondence in the majority of the basic sections", "correspondence in all basic sections", "complete correspondence".

Table 3.6 Results of an expert evaluation

Tutorials	Expert					
	1	2	3	4	5	6
1	6	5	6	5	5	6
2	4	4	3	3	4	4
3	1	1	2	1	1	1
4	6	5	5	5	5	6
5	2	2	3	2	3	2
6	5	4	5	4	4	5
7	1	1	1	2	2	1
8	6	6	6	6	6	6
9	5	4	5	5	5	5
10	3	3	3	3	3	3
11	3	2	3	2	2	3
12	5	5	5	5	5	5
13	6	6	6	6	6	6
14	6	5	6	4	5	6
15	3	3	3	1	3	3
16	4	4	4	4	4	4

Table 3.6 (*continued*)

17	4	3	4	3	2	4
18	6	6	6	6	5	6
19	6	6	6	6	5	6
20	4	3	3	3	2	4

Table 3.7 Relative results of expert evaluations

Tutorials	Expert					
	1	2	3	4	5	6
1	0,1	0,1	0,05	0,1	0,05	0,1
2	0,05	0,1	0,05	0,15	0,2	0,05
3	0,15	0,2	0,3	0,2	0,15	0,15
4	0,2	0,2	0,1	0,15	0,15	0,2
5	0,15	0,25	0,2	0,2	0,35	0,15
6	0,35	0,15	0,3	0,2	0,1	0,35

Expert evaluations in the form of numbers of the verbal scale levels are given in Table 3.6.

Relative results of expert evaluations within the limits of each level are summarized in Table 3.7. Number of a line corresponds to number of the rating scale level.

Six COSS's X_i; $i = \overline{1,6}$ are constructed, which formalize criteria of expert evaluations of courseware correspondence to curricula. While constructing, data of Table 3.7 and a method described in § 2.2 were used. Parameters of membership functions $\mu_{il}(x)$; $i = \overline{1,6}$; $l = \overline{1,6}$ of term-sets COSS are summarized in Table 3.8, and parameters of membership functions of the formalized expert results — in Table 3.9.

Indexes of the general consistency of results are the following: the additive index is equal to 0.218, the multiplicative index is equal to 0, that indicates availability of significantly differing results. For comparison, the concordation coefficient calculated within the expert ranking limits is equal to 0.07.

Elements of a matrix of a pairwise similarity of the formalized expert results are presented below.

Elements of a matrix of a pairwise similarity of the formalized expert results

1	0,57	0,75	0,54	0,48	1
0,57	1	0,61	0,84	0,79	0,57
0,75	0,61	1	0,59	0,59	0,75
0,54	0,84	0,59	1	0,75	0,54
0,48	0,79	0,59	0,75	1	0,48
1	0,57	0,75	0,54	0,48	1

Since the matrix obtained is not transitive, its transitive closure is determined and by that the similarity relation $\hat{R}_1$ is revealed. Using the decomposition theorem, $\hat{R}_1$ are decomposed to equivalence relations and the obtained results presented below.

Fuzzy clusterization of expert results versus similarity relation $\hat{R}_1$

Confidence level	Cluster
0,61	{1,2,3,4,5,6}
0,75	{1,3,6}, {2,4,5}
0,79	{1,6}, {2,4,5},{3}
0,84	{1,6}, {2,4,},{3},{5}
1	{1,6}, {2}, {3}, {4}, {5}

Apparently, 1st and 6th expert results completely coincide, 2nd, 4th and 5th expert results are similar, but they differ from the results of 1st and 6th experts.

Table 3.8 Parameters of membership functions of term-sets of the formalized expert approaches

Membership function	X_1	X_2	X_3
μ_{i1}	(0;0,075;0;0,05)	(0;0,05;0;0,1)	(0;0,025;0;0,05)
μ_{i2}	(0,125;0,05;0,05)	(0,15;0,1;0,1)	(0,075;0,05;0,05)
μ_{i3}	(0,175;0,225;0,05;0,15)	(0,25;0,3;0,1;0,2)	(0,125;0,35;0,05;0,1)
μ_{i4}	(0,375;0,425;0,15;0,15)	(0,5;0,2;0,2)	(0,45;0,1;0,1)
μ_{i5}	(0,575;0,15;0,15)	(0,7;0,775;0,2;0,15)	(0,55;0,6;0,1;0,2)
μ_{i6}	(0,725;1;0,15;0)	(0,925;1;0,15;0)	(0,8;1;0,2;0)

X_4	X_5	X_6
(0;0,05;0;0,1)	(0;0,025;0;0,05)	(0;0,075;0;0,05)
(0,15;0,175;0,1;0,15)	(0,075;0,175;0,05;0,15)	(0,125;0,05;0,05)
(0,325;0,375;0,15;0,15)	(0,325;0,15;0,15)	(0,175;0,225;0,05;0,15)
(0,525;0,15;0,15)	(0,475;0,15;0,15)	(0,375;0,425;0,15;0,15)
(0,675;0,7;0,15;0,2)	(0,625;0,85;0,15;0,1)	(0,575;0,15;0,15)
(0,9;1;0,2;0)	(0,95;1;0,1;0)	(0,725;1;0,15;0)

Table 3.9 Parameters of membership functions of the formalized results of experts

Membership function	M_1	M_2	M_3
μ_i^1	(0,725;1;0,15;0)	(0,7;0,775;0,2;0,15)	(0,8;1;0,2;0)
μ_i^2	(0,375;0,425;0,15;0,15)	(0,5;0,2;0,2)	(0,125;0,35;0,05;0,1)
μ_i^3	(0;0,075;0;0,05)	(0;0,05;0;0,1)	(0,075;0,05;0,05)
μ_i^4	(0,725;1;0,15;0)	(0,7;0,775;0,2;0,15)	(0,55;0,6;0,1;0,2)
μ_i^5	(0,125;0,05;0,05)	(0,15;0,1;0,1)	(0,125;0,35;0,05;0,1)
μ_i^6	(0,575;0,15;0,15)	(0,5;0,2;0,2)	(0,55;0,6;0,1;0,2)
μ_i^7	(0;0,075;0;0,05)	(0;0,05;0;0,1)	(0;0,025;0;0,05)
μ_i^8	(0,725;1;0,15;0)	(0,925;1;0,15;0)	(0,8;1;0,2;0)
μ_i^9	(0,575;0,15;0,15)	(0,5;0,2;0,2)	(0,55;0,6;0,1;0,2)
μ_i^{10}	(0,175;0,225;0,05;0,15)	(0,25;0,3;0,1;0,2)	(0,125;0,35;0,05;0,1)
μ_i^{11}	(0,175;0,225;0,05;0,15)	(0,15;0,1;0,1)	(0,125;0,35;0,05;0,1)
μ_i^{12}	(0,575;0,15;0,15)	(0,7;0,775;0,2;0,15)	(0,55;0,6;0,1;0,2)
μ_i^{13}	(0,725;1;0,15;0)	(0,925,1;0,15;0)	(0,8;1;0,2;0)
μ_i^{14}	(0,725;1;0,15;0)	(0,7;0,775;0,2;0,15)	(0,8;1;0,2;0)
μ_i^{15}	(0,175;0,225;0,05;0,15)	(0,25;0,3;0,1;0,2)	(0,125;0,35;0,05;0,1)
μ_i^{16}	(0,375;0,425;0,15;0,15)	(0,5;0,2;0,2)	(0,45;0,1;0,1)
μ_i^{17}	(0,375;0,425;0,15;0,15)	(0,25;0,3;0,1;0,2)	(0,45;0,1;0,1)
μ_i^{18}	(0,725;1;0,15;0)	(0,925;1;0,15;0)	(0,8;1;0,2;0)
μ_i^{19}	(0,725;1;0,15;0)	(0,7;0,775;0,2;0,15)	(0,8;1;0,2;0)
μ_i^{20}	(0,375;0,425;0,15;0,15)	(0,25;0,3;0,1;0,2)	(0,125;0,35;0,05;0,1)

M_4	M_5	M_6
(0,675;0,7;0,15;0,2)	(0,625;0,85;0,15;0,1)	(0,725;1;0,15;0)
(0,35;0,375;0,2;0,15)	(0,475;0,15;0,15)	(0,375;0,425;0,15;0,15)
(0;0,05;0;0,1)	(0;0,025;0;0,15)	(0;0,075;0;0,05)

Table 3.9 (*continued*)

(0,675;0,7;0,15;0,2)	(0,625;0,85;0,15;0,1)	(0,725;1;0,15;0)
(0,15;0,1;0,2)	(0,325;0,15;0,15)	(0,125;0,05;0,05)
(0,525;0,15;0,15)	(0,475;0,15;0,15)	(0,575;0,15;0,15)
(0,15;0,1;0,2)	(0,075;0,175;0,05; 0,15)	(0;0,075;0;0,05)
(0,9;1;0,2;0)	(0;95;1;0,1;0)	(0,725;1;0,15;0)
(0,675;0,7;0,15;0,2)	(0,625;0,85;0,15;0,1)	(0,575;0,15;0,15)
(0,35;0,375;0,2;0,15)	(0,325;0,15;0,15)	(0,175;0,225;0,05;0,15)
(0,15;0,1;0,2)	(0,075;0,175;0,05;0,15)	(0,175;0,225;0,05;0,15)
(0,675;0,7;0,15;0,2)	(0,625;0,85;0,15;0,1)	(0,575;0,15;0,15)
(0,9;1;0,2;0)	(0;95;1;0,1;0)	(0,725;1;0,15;0)
(0,525;0,15;0,15)	(0,625;0,85;0,15;0,1)	(0,725;1;0,15;0)
(0;0,05;0;0,1)	(0,325;0,15;0,15)	(0,175;0,225;0,05;0,15)
(0,525;0,15;0,15)	(0,475;0,15;0,15)	(0,375;0,425;0,15;0,15)
(0,35;0,375;0,2;0,15)	(0,075;0,175;0,05;0,15)	(0,375;0,425;0,15;0,15)
(0,9;1;0,2;0)	(0,625;0,85;0,15;0,1)	(0,725;1;0,15;0)
(0,9;1;0,2;0)	(0,625;0,85;0,15;0,1)	(0,725;1;0,15;0)
(0,35;0,375;0,2;0,15)	(0,075;0,175;0,05;0,15)	(0,375;0,425;0,15;0,15)

Results of the 3rd expert differ from results of all experts, but they are closer to results of the 1st and 6th experts.

Based on the research carried out, the conclusion is drawn that the results obtained by the third expert essentially differs from all results. It is confirmed by the fact that the system of results without his results included has the following general consistency indexes: additive — 0.266, multiplicative — 0.252. Considering that results of experts are divided into two similar groups, indexes of consistency for two subsystems {2, 4, 5}, and {1, 3, 6} are obtained. The first subsystem has additive and multiplicative indexes equal to 0.522 and 0.5204, and the second subsystem — 0.503 and 0, accordingly.

Example 3.4. The analysis of results of an expert evaluation of "large-leaved linden" species state. Three experts estimated the state of 40 of "large-leaved linden" plants in a center of Moscow. For this purpose the verbal scale with seven levels X_l $\left(l = \overline{1,7}\right)$ was used: "old dead standing tree", "recent dead standing trees", "partial mortal (drying)", "very weakened tree", "mean weakened tree", "moderately weakened tree", "vigorous tree without weakening signs" [148—149]. The obtained data are provided in Table 3.10.

Let us denote with a_l^i $\left(l = \overline{1,3}; l = \overline{1,7}\right)$ a relative amount of the trees referred by i-th expert to l-th level of a state. The data obtained are shown below.

Relative amounts of trees of each state level of a condition according to three experts

$$a_1^1\ a_2^1\ a_3^1\ a_4^1\ a_5^1\ a_6^1\ a_7^1\quad a_1^2\ a_2^2\ a_3^2\ a_4^2\ a_5^2\ a_6^2\ a_7^2\quad a_1^3\ a_2^3\ a_3^3\ a_4^3\ a_5^3\ a_6^3\ a_7^3$$

1	1	1	3	1	17	7	1	0	1	1	3	17	3	1	1	1	1	13	19	1
40	40	40	40	4	40	40	40	20	20	10	40	20	40	40	40	40	40	40	10	

Based on these data and the method described in§ 2.2, COSS's $X_i\ (i = \overline{1,3})$ are constructed, they are formalizations of expert approaches to the plants state evaluation.

Table 3.10 Data of evaluations by three experts of forty "large-leaved linden" trees state

No. of item	1th expert							2nd expert							3rd expert						
	1	2	3	4	5	6	7	1	2	3	4	5	6	7	1	2	3	4	5	6	7
1							+							+	+						
2							+		+						+						
3							+							+	+						
4					+								+		+						
5					+								+		+						
6						+							+		+						
7					+								+		+						
8						+							+		+						
9							+							+	+						
10						+							+		+						
11					+								+		+						
12					+								+		+						
13					+								+		+						
14					+								+		+						
15						+								+	+						
16					+								+		+						
17					+								+		+						
18						+								+	+						
19					+								+		+						
20					+								+		+						
21				+									+		+						
22						+							+		+						
23				+									+		+						
24				+										+	+						
25					+								+		+						

Table 3.10 (*continued*)

26		+	+	+
27		+	+	+
28	+	+	+
29		+	+	+
30		+	+	+
31	+	+	+
32		+	+	+
33		+	+	+
34		+	+	+
35		+	+	+
36	+	+	+
37		+	+	+
38	+	+	+
39		+	+	+
40		+	+	+

Table 3.11 Parameters of membership functions of the formalized expert approaches to the plants state evaluation

Membership function	X_1	X_2	X_3
μ_{i1}	(0; 0,013; 0; 0,025)	(0; 0,025; 0; 0)	(0; 0,013; 0; 0,025)
μ_{i2}	(0,038; 0,038; 0,025; 0,025)	(0,025; 0,025; 0; 0)	(0,038; 0,038; 0,025; 0,025;
μ_{i3}	(0,063; 0,063; 0,025; 0,025)	(0,025; 0,050; 0; 0,050)	(0,063; 0,063; 0,025; 0,025
μ_{i4}	(0,088; 0,113; 0,025; 0,075)	(0,100; 0,100; 0,050; 0,050)	(0,088; 0,088; 0,025; 0,025
μ_{i5}	(0,188; 0,275; 0,075; 0,250)	(0,150; 0,275; 0,050; 0,300)	(0,113; 0,263; 0,025; 0,325
μ_{i6}	(0,525; 0,738; 0,250; 0,175)	(0,575; 0,775; 0,300; 0,150)	(0,588; 0,850; 0,325; 0,100
μ_{i7}	(0,913; 1,000; 0,175; 0)	(0,925; 1,000; 0,150; 0)	(0,950; 1,000; 0,100; 0)

Parameters of membership function of term-sets of the formalized approaches are summarized in Table 3.11.

Having substituted a sign "+" in Table 3.10 with the parameters of corresponding membership functions, we obtain the formalized expert results.

The additive index (3.24) of general consistency of the formalized evaluation results obtained by three experts is equal to 0.53, that is evidence of competence of the experts.

Elements of a matrix of pairwise similarity and pairwise consistency of expert results are presented below.

Elements of a matrix of pairwise similarity

$$
\begin{array}{cc}
1 & 0,857 \\
0,857 & 1 \\
0,823 & 0,863
\end{array}
\quad
\begin{array}{c}
0,823\ 0,863 \\
1
\end{array}
$$

Elements of a matrix of pairwise consistency

$$
\begin{array}{ccc}
1 & 0,684 & 0,635 \\
0,684 & 1 & 0,685 \\
0.635 & 0,685 & 1
\end{array}
$$

The matrix of pairwise similarity allows to conclude that results of the 2nd and 3rd experts are of the greatest similarity. It means that they not only refer the most representative part of the considered group to the same terms, but also representatives within the terms mainly coincide. The matrix of pairwise consistency allows to conclude that the greatest consistency is for results of 2nd and 3rd experts. These results are similar not only within the limits of a great bulk of plants, but also are uniform for all representatives of the group considered.

Matrixes of pairwise similarity and pairwise consistency are matrixes of fuzzy relations of similarity R_1 and R_2 over set of the formalized expert evaluations, accordingly. To construct fuzzy relations of similarity, transitive closures R_1 and $R_2 - \hat{R}_1$ and $\hat{R}_2$ are to be defined. Outcomes of the fuzzy cluster analysis of the formalized expert results under similarity relation $\hat{R}_1$ and $\hat{R}_2$ are presented below.

Fuzzy clusterization of expert evaluations

α -level	Cluster
Similarity $\hat{R}_1$	
0,857	{1,2,3}
0863	{1},{2,3}
1	{1},{2},{3}
Similarity $\hat{R}_2$	
0,684	{1,2,3}
0,685	{1},{2,3}
1	{1},{2},{3}

The obtained results allow concluding that the plants state evaluation results obtained by 2nd and 3rd experts are of the most similarity.

Example 3.5. Clusterization of the plant species states on the basis of expert evaluations. This problem is directly connected with problems of definition of plant species which are the most adapted for severe ecological conditions of a big city and problems of planning of various measures to be performed at amenity planting objects. Usually it is insufficient just to select the best (in a certain sense) specie of plants for specific growth conditions, it is necessary to carry out clusterization of these species (on the basis of paired comparison) to define the similar species.

One of prime ecological monitoring problems is the problem of an expert evaluation of greenery states [149], it consists in visual inspection of the greenery and reference of individual specimens to one of the verbal scale levels. For the purpose of such evaluation the scale described in the example 3.1 was used.

In 1997-2001, in Moscow experts inspected seventeen species of woody plants and brushwood growing in severe ecological conditions of the Boulevard Ring avenue within the municipal program of greenery monitoring (4084 plants were inspected) [149]. One of the inspection purposes was making a solution on prospectivity of use of these species for city garden.

Table 3.12 Relative results of the greenery state evaluations

№	Specie	a_{i1}^1	a_{i2}^1	a_{i3}^1	a_{i4}^1	a_{i5}^1	a_{i6}^1	a_{i7}^1
1	European white birch	0,090	0,180	0,000	0,180	0,180	0,370	0,000
2	European hawthorn	0,000	0,000	0,040	0,149	0,400	0,260	0,151
3	European white elm	0,000	0,005	0,014	0,057	0,220	0,650	0,054
4	Witch elm	0,023	0,000	0,018	0,123	0,220	0,474	0,142
5	Single-seed hawthorn	0,000	0,000	0,000	0,040	0,240	0,680	0,040
6	Cotoneaster	0,000	0,000	0,000	0,083	0,125	0,670	0,122
7	Norway maple	0,023	0,008	0,031	0,160	0,183	0,400	0,195
8	Tatarian maple	0,000	0,052	0,069	0,069	0,276	0,413	0,121
9	Canadian maple	0,000	0,000	0,014	0,220	0,324	0,408	0,034
10	Large-leaved linden	0,002	0,009	0,022	0,087	0,274	0,426	0,180
11	Little-leaved linden	0,005	0,008	0,050	0,140	0,336	0,392	0,069
12	Hungarian lilac	0,020	0,010	0,040	0,131	0,354	0,374	0,071
13	Common lilac	0,000	0,005	0,018	0,074	0,310	0,490	0,103
14	Cottonwood	0,023	0,058	0,058	0,058	0,151	0,477	0,175
15	Rough-bark poplar	0,039	0,031	0,035	0,117	0,190	0,432	0,156
16	European ash	0,000	0,000	0,077	0,000	0,307	0,462	0,154
17	Black ash	0,003	0,002	0,012	0,161	0,326	0,447	0,049

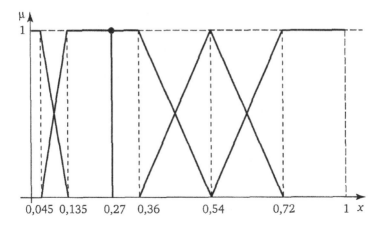

Fig. 3.1 Membership functions of COSS term-set X_1 "a state of European white birch plants in the Boulevard Ring avenue"

The abundance of plants of i-th specie in the greenery of the Boulevard Ring avenue which was referred by experts to l-th level, is denoted with a_{il}^1 $\left(i = \overline{1,17}; l = \overline{1,7}\right)$. The data obtained are summarized in Table 3.12.

Based on data of Table 3.12, seventeen COSS's Y_i $\left(i = \overline{1,17}\right)$ with term-sets X_l $\left(i = \overline{1,7}\right)$ were constructed using the method of § 2.2. The fuzzy numbers corresponding to term-sets X_l are T-numbers or normal triangular numbers with membership functions $\mu_{il}(x)$, accordingly (Table 3.13).

As an example, graphs of membership functions of COSS X_1 term-set are shown in Fig. 3.1.

Using definitions of §3.1, paired comparative characteristics of states of various plant species were defined. Elements of the pairwise similarity matrix are given in Table 3.14.

Elements of the pairwise consistency matrix are summarized in Table 3.15.

The additive index (3.4) of general consistency of states of all plant species is equal to 0.11 thus evidencing that states of plant species considerably differ.

Fuzzy clusterization for 17 species of woody plants and brushwood, the states of which are formalized above, was made. Fuzzy binary relations of similarity over the set of the formalized states are constructed on the basis of matrixes of pairwise similarity and pairwise consistency the elements which are provided in Table 3.14 and Table3.15, accordingly.

Table 3.13 Parameters of COSS membership functions

μ	European white birch	European hawthorn	European white elm
μ_{i1}	0,000 0,045 0,000 0,090	0,000 0,000 0,000 0,000	0,000 0,000 0,000 0,000
μ_{i2}	0,135 0,270 0,090 0,000	0,000 0,000 0,000 0,000	0,000 0,003 0,000 0,005
μ_{i3}	0,270 0,270 0,000 0,000	0,000 0,020 0,000 0,040	0,008 0,012 0,005 0,014
μ_{i4}	0,270 0,360 0,000 0,180	0,060 0,115 0,040 0,149	0,026 0,048 0,014 0,057
μ_{i5}	0,540 0,540 0,180 0,180	0,264 0,459 0,149 0,260	0,105 0,186 0,057 0,220
μ_{i6}	0,720 1,000 0,180 0,000	0,719 0,774 0,260 0,151	0,406 0,919 0,220 0,054
μ_{i7}	1,000 1,000 0,000 0,000	0,925 1,000 0,151 0,000	0,973 1,000 0,054 0,000

μ	Witch elm	Single-seed hawthorn	Cotoneaster
μ_{i1}	0,000 0,023 0,000 0,000	0,000 0,000 0,000 0,000	0,000 0,000 0,000 0,000
μ_{i2}	0,023 0,023 0,000 0,000	0,000 0,000 0,000 0,000	0,000 0,000 0,000 0,000
μ_{i3}	0,023 0,032 0,000 0,018	0,000 0,000 0,000 0,000	0,000 0,000 0,000 0,000
μ_{i4}	0,050 0,103 0,018 0,123	0,000 0,020 0,000 0,040	0,000 0,042 0,000 0,083
μ_{i5}	0,226 0,274 0,123 0,220	0,060 0,160 0,040 0,240	0,125 0,146 0,083 0,125
μ_{i6}	0,494 0,787 0,220 0,142	0,400 0,940 0,240 0,040	0,271 0,817 0,125 0,122
μ_{i7}	0,929 1,000 0,142 0,000	0,980 1,000 0,040 0,000	0,939 1,000 0,122 0,000

μ	Norway maple	Tatarian maple	Canadian maple
μ_{i1}	0,000 0,019 0,000 0,008	0,000 0,000 0,000 0,000	0,000 0,000 0,000 0,000
μ_{i2}	0,027 0,027 0,008 0,008	0,000 0,026 0,000 0,052	0,000 0,000 0,000 0,000
μ_{i3}	0,035 0,047 0,008 0,031	0,078 0,087 0,052 0,069	0,000 0,007 0,000 0,014
μ_{i4}	0,078 0,142 0,031 0,160	0,156 0,156 0,069 0,069	0,021 0,124 0,014 0,220
μ_{i5}	0,302 0,314 0,160 0,183	0,225 0,328 0,069 0,276	0,344 0,396 0,220 0,324
μ_{i6}	0,497 0,708 0,183 0,195	0,604 0,819 0,276 0,121	0,720 0,949 0,324 0,034
μ_{i7}	0,903 1,000 0,195 0,000	0,940 1,000 0,121 0,000	0,983 1,000 0,034 0,000

Table 3.13 (*continued*)

μ	Large-leaved linden	Little-leaved linden	Hungarian lilac
μ_{i1}	0,000 0,001 0,000 0,002	0,000 0,003 0,000 0,005	0,000 0,015 0,000 0,010
μ_{i2}	0,003 0,007 0,002 0,009	0,008 0,009 0,005 0,008	0,025 0,025 0,010 0,010
μ_{i3}	0,016 0,022 0,009 0,022	0,017 0,038 0,008 0,050	0,035 0,050 0,010 0,040
μ_{i4}	0,044 0,077 0,022 0,087	0,088 0,133 0,050 0,140	0,090 0,136 0,040 0,131
μ_{i5}	0,164 0,257 0,087 0,274	0,273 0,371 0,140 0,336	0,267 0,378 0,131 0,354
μ_{i6}	0,531 0,730 0,274 0,180	0,707 0,897 0,336 0,069	0,732 0,894 0,354 0,071
μ_{i7}	0,910 1,000 0,180 0,000	0,966 1,000 0,069 0,000	0,965 1,000 0,071 0,000

μ	Common lilac	Cottonwood	Rough-bark poplar
μ_{i1}	0,000 0,000 0,000 0,000	0,000 0,012 0,000 0,023	0,000 0,024 0,000 0,031
μ_{i2}	0,000 0,003 0,000 0,005	0,035 0,052 0,023 0,058	0,055 0,055 0,031 0,031
μ_{i3}	0,008 0,014 0,005 0,018	0,110 0,110 0,058 0,058	0,086 0,088 0,031 0,035
μ_{i4}	0,032 0,060 0,018 0,074	0,168 0,168 0,058 0,058	0,123 0,164 0,035 0,117
μ_{i5}	0,134 0,252 0,074 0,310	0,226 0,273 0,058 0,151	0,281 0,317 0,117 0,190
μ_{i6}	0,562 0,846 0,310 0,103	0,424 0,738 0,151 0,175	0,507 0,766 0,190 0,156
μ_{i7}	0,949 1,000 0,103 0,000	0,913 1,000 0,175 0,000	0,922 1,000 0,156 0,000

μ	European ash				Black ash			
μ_{i1}	0,000	0,000	0,000	0,000	0,000	0,002	0,000	0,002
μ_{i2}	0,000	0,000	0,000	0,000	0,004	0,004	0,002	0,002
μ_{i3}	0,000	0,077	0,000	0,000	0,006	0,011	0,002	0,012
μ_{i4}	0,077	0,077	0,000	0,000	0,023	0,098	0,012	0,161
μ_{i5}	0,077	0,231	0,000	0,307	0,259	0,341	0,161	0,326
μ_{i6}	0,538	0,769	0,307	0,154	0,667	0,927	0,326	0,049
μ_{i7}	0,923	1,000	0,154	0,000	0,976	1,000	0,049	0,000

The pairwise similarity matrix can be recommended for use in the case when an expert in charge of the analysis of greenery state, is interested in the similarity of a great bulk of representatives of some specific species to each other. In this case contributions of the least representative levels will be amortized by contributions of the most representative ones, due to their similarity.

The pairwise consistency matrix can be recommended for use in the case when an expert is interested in the complete-scaled picture considering uniform contributions of all levels, including the smallest ones, due to their similarity.

Table 3.14 Elements of the pairwise similarity matrix of plant species states

1,000 0,364 0,326 0,282 0,352 0,248 0,227 0,318 0,494 0,240 0,427 0,447 0,318 0,238 0,288 0,247 0,424

0,364 1,000 0,482 0,743 0,436 0,491 0,700 0,717 0,794 0,693 0,820 0,824 0,659 0,593 0,700 0,645 0,775

0,326 0,482 1,000 0,725 0,918 0,819 0,576 0,582 0,581 0,721 0,581 0,558 0,817 0,644 0,594 0,745 0,689

0,282 0,743 0,725 1,000 0,655 0,707 0,844 0,773 0,624 0,900 0,706 0,689 0,854 0,794 0,826 0,830 0,760

0,352 0,436 0,918 0,655 1,000 0,803 0,537 0,550 0,558 0,640 0,531 0,519 0,734 0,620 0,571 0,706 0,626

0,248 0,491 0,819 0,707 0,803 1,000 0,590 0,573 0,461 0,688 0,474 0,462 0,747 0,641 0,604 0,701 0,543

0,227 0,700 0,576 0,844 0,537 0,590 1,000 0,746 0,619 0,825 0,707 0,697 0,705 0,789 0,875 0,750 0,667

0,318 0,717 0,582 0,773 0,550 0,573 0,746 1,000 0,651 0,710 0,764 0,766 0,746 0,765 0,830 0,720 0,755

0,494 0,794 0,581 0,624 0,558 0,461 0,619 0,651 1,000 0,556 0,857 0,861 0,639 0,483 0,620 0,498 0,868

0,240 0,693 0,721 0,900 0,640 0,688 0,825 0,710 0,556 1,000 0,631 0,606 0,877 0,774 0,777 0,869 0,693

0,427 0,820 0,581 0,706 0,531 0,474 0,707 0,764 0,857 0,631 1,000 0,963 0,683 0,567 0,691 0,610 0,871

0,447 0,824 0,558 0,689 0,519 0,462 0,697 0,766 0,861 0,606 0,963 1,000 0,661 0,571 0,686 0,594 0,852

0,318 0,659 0,817 0,854 0,734 0,747 0,705 0,746 0,639 0,877 0,683 0,661 1,000 0,683 0,725 0,847 0,763

0,238 0,593 0,644 0,794 0,620 0,641 0,789 0,765 0,483 0,774 0,567 0,571 0,683 1,000 0,843 0,753 0,574

0,288 0,700 0,594 0,826 0,571 0,604 0,875 0,830 0,620 0,777 0,691 0,686 0,725 0,843 1,000 0,771 0,675

0,247 0,645 0,745 0,830 0,706 0,701 0,750 0,720 0,498 0,869 0,610 0,594 0,847 0,753 0,771 1,000 0,611

0,424 0,775 0,689 0,760 0,626 0,543 0,667 0,755 0,868 0,693 0,871 0,852 0,763 0,574 0,675 0,611 1,000

Table 3.15 Elements of the pairwise consistency matrix of plant species states

1,000 0,123 0,208 0,240 0,066 0,045 0,091 0,093 0,162 0,202 0,143 0,173 0,220 0,091 0,130 0,060 0,272

0,123 1,000 0,313 0,391 0,406 0,501 0,348 0,494 0,668 0,331 0,389 0,398 0,426 0,287 0,355 0,635 0,352

0,208 0,313 1,000 0,398 0,502 0,494 0,183 0,351 0,381 0,420 0,247 0,234 0,794 0,202 0,181 0,389 0,462

0,240 0,391 0,398 1,000 0,192 0,287 0,678 0,375 0,244 0,738 0,484 0,566 0,525 0,486 0,484 0,357 0,637

0,066 0,406 0,502 0,192 1,000 0,733 0,149 0,311 0,513 0,187 0,195 0,187 0,374 0,176 0,159 0,504 0,250

0,045 0,501 0,494 0,287 0,733 1,000 0,207 0,386 0,439 0,289 0,181 0,177 0,499 0,217 0,213 0,548 0,202

0,091 0,348 0,183 0,678 0,149 0,207 1,000 0,373 0,261 0,544 0,577 0,690 0,286 0,511 0,566 0,363 0,437

0,093 0,494 0,351 0,375 0,311 0,386 0,373 1,000 0,406 0,305 0,386 0,395 0,483 0,494 0,517 0,462 0,333

0,162 0,668 0,381 0,244 0,513 0,439 0,261 0,406 1,000 0,205 0,408 0,405 0,382 0,170 0,244 0,462 0,427

0,202 0,331 0,420 0,738 0,187 0,289 0,544 0,305 0,205 1,000 0,442 0,392 0,562 0,368 0,363 0,376 0,659

0,143 0,389 0,247 0,484 0,195 0,181 0,577 0,386 0,408 0,442 1,000 0,804 0,290 0,275 0,346 0,289 0,660

0,173 0,398 0,234 0,566 0,187 0,177 0,690 0,395 0,405 0,392 0,804 1,000 0,283 0,366 0,404 0,265 0,580

Table 3.15 (*continued*)

0,220 0,426 0,794 0,525 0,374 0,499 0,286 0,483 0,382 0,562 0,290 0,283 1,000 0,260 0,287 0,481 0,461
0,091 0,287 0,202 0,486 0,176 0,217 0,511 0,494 0,170 0,368 0,275 0,366 0,260 1,000 0,620 0,324 0,234
0,130 0,355 0,181 0,484 0,159 0,213 0,566 0,517 0,244 0,363 0,346 0,404 0,287 0,620 1,000 0,352 0,280
0,060 0,635 0,389 0,357 0,504 0,548 0,363 0,462 0,462 0,376 0,289 0,265 0,481 0,324 0,352 1,000 0,200
0,272 0,352 0,462 0,637 0,250 0,202 0,437 0,333 0,427 0,659 0,660 0,580 0,461 0,234 0,280 0,200 1,000

Table 3.16 Elements of the similarity relation $\hat{R}_1$ matrix

1,000 0,494 0,494 0,494 0,494 0,494 0,494 0,494 0,494 0,494 0,494 0,494 0,494 0,494 0,494 0,494
0,494 1,000 0,766 0,766 0,766 0,766 0,766 0,766 0,824 0,766 0,824 0,824 0,766 0,766 0,766 0,766
0,494 0,766 1,000 0,817 0,918 0,819 0,817 0,817 0,766 0,817 0,766 0,766 0,817 0,817 0,817 0,817
0,494 0,766 0,817 1,000 0,817 0,817 0,844 0,830 0,766 0,900 0,766 0,766 0,877 0,843 0,844 0,869
0,494 0,766 0,918 0,817 1,000 0,819 0,817 0,817 0,766 0,817 0,766 0,766 0,817 0,817 0,817 0,817
0,494 0,766 0,819 0,817 0,819 1,000 0,817 0,817 0,766 0,817 0,766 0,766 0,817 0,817 0,817 0,817
0,494 0,766 0,817 0,844 0,817 0,817 1,000 0,830 0,766 0,844 0,766 0,766 0,844 0,843 0,875 0,844
0,494 0,766 0,817 0,830 0,817 0,817 0,830 1,000 0,766 0,830 0,766 0,766 0,830 0,830 0,830 0,830
0,494 0,824 0,766 0,766 0,766 0,766 0,766 0,766 1,000 0,766 0,868 0,868 0,766 0,766 0,766 0,766
0,494 0,766 0,817 0,900 0,817 0,817 0,844 0,830 0,766 1,000 0,766 0,766 0,877 0,843 0,844 0,869
0,494 0,824 0,766 0,766 0,766 0,766 0,766 0,766 0,868 0,766 1,000 0,963 0,766 0,766 0,766 0,766
0,494 0,824 0,766 0,766 0,766 0,766 0,766 0,766 0,868 0,766 0,963 1,000 0,766 0,766 0,766 0,766
0,494 0,766 0,817 0,877 0,817 0,817 0,844 0,830 0,766 0,877 0,766 0,766 1,000 0,843 0,844 0,869
0,494 0,766 0,817 0,843 0,817 0,817 0,843 0,830 0,766 0,843 0,766 0,766 0,843 1,000 0,843 0,843
0,494 0,766 0,817 0,844 0,817 0,817 0,875 0,830 0,766 0,844 0,766 0,766 0,844 0,843 1,000 0,844
0,494 0,766 0,817 0,869 0,817 0,817 0,844 0,830 0,766 0,869 0,766 0,766 0,869 0,843 0,844 1,000
0,494 0,824 0,766 0,766 0,766 0,766 0,766 0,766 0,868 0,766 0,871 0,871 0,766 0,766 0,766 0,766

Table 3.17 Elements of the similarity relation $\hat{R}_2$ matrix

1,000 0,272 0,272 0,272 0,272 0,272 0,272 0,272 0,272 0,272 0,272 0,272 0,272 0,272 0,272 0,272 0,272
0,272 1,000 0,502 0,502 0,548 0,548 0,502 0,502 0,668 0,502 0,502 0,502 0,502 0,502 0,502 0,635 0,502
0,272 0,502 1,000 0,562 0,502 0,502 0,562 0,517 0,502 0,562 0,562 0,562 0,794 0,562 0,562 0,502 0,562
0,272 0,502 0,562 1,000 0,502 0,502 0,678 0,517 0,502 0,738 0,678 0,678 0,562 0,566 0,566 0,502 0,660
0,272 0,548 0,502 0,502 1,000 0,733 0,502 0,502 0,548 0,502 0,502 0,502 0,502 0,502 0,502 0,548 0,502
0,272 0,548 0,502 0,502 0,733 1,000 0,502 0,502 0,548 0,502 0,502 0,502 0,502 0,502 0,502 0,548 0,502
0,272 0,502 0,562 0,678 0,502 0,502 1,000 0,517 0,502 0,678 0,690 0,690 0,562 0,566 0,566 0,502 0,660
0,272 0,502 0,517 0,517 0,502 0,502 0,517 1,000 0,502 0,517 0,517 0,517 0,517 0,517 0,517 0,502 0,517
0,272 0,668 0,502 0,502 0,548 0,548 0,502 0,502 1,000 0,502 0,502 0,502 0,502 0,502 0,502 0,635 0,502
0,272 0,502 0,562 0,738 0,502 0,502 0,678 0,517 0,502 1,000 0,678 0,678 0,562 0,566 0,566 0,502 0,660
0,272 0,502 0,562 0,678 0,502 0,502 0,690 0,517 0,502 0,678 1,000 0,804 0,562 0,566 0,566 0,502 0,660
0,272 0,502 0,562 0,678 0,502 0,502 0,690 0,517 0,502 0,678 0,804 1,000 0,562 0,566 0,566 0,502 0,660

Table 3.17 (*continued*)

0,272	0,502	0,794	0,562	0,502	0,502	0,562	0,517	0,502	0,562	0,562	0,562	1,000	0,562	0,562	0,502	0,562
0,272	0,502	0,562	0,566	0,502	0,502	0,566	0,517	0,502	0,566	0,566	0,566	0,562	1,000	0,620	0,502	0,566
0,272	0,502	0,562	0,566	0,502	0,502	0,566	0,517	0,502	0,566	0,566	0,566	0,562	0,620	1,000	0,502	0,566
0,272	0,635	0,502	0,502	0,548	0,548	0,502	0,502	0,635	0,502	0,502	0,502	0,502	0,502	0,502	1,000	0,502
0,272	0,502	0,562	0,660	0,502	0,502	0,660	0,517	0,502	0,660	0,660	0,660	0,562	0,566	0,566	0,502	1,000

As one can see from Table 3.14 and Table 3.15, fuzzy relations R_1 and R_2 defined by matrixes of pairwise similarity and pairwise consistency, accordingly, are not transitive, therefore their transitive closures $\hat{R}_1$ and $\hat{R}_2$ which are similarity relations, are constructed. Elements of relations matrixes $\hat{R}_1$ and $\hat{R}_2$ are provided in Table 3.16 and Table 3.17, accordingly.

Fuzzinesses of similarity relations are $\hat{R}_1 = 0,433$, $\hat{R}_2 = 0,805$. As known [15], fuzziness degree of fuzzy relation R

$$\xi(R) = \frac{2}{n \cdot m} \sum_{i,j} \left| \mu_R(x_i, y_j) - \tilde{r}_{ij} \right|,$$

where $\tilde{r}_{ij} = \begin{cases} 0, & \mu_R(x_i, y_j) \le 0,5; \\ 1, & \mu_R(x_i, y_j) > 0,5. \end{cases}$

Owing to higher fuzziness of similarity relation $\hat{R}_2$, the decision to use the similarity relation $\hat{R}_1$ for the further analysis was made. Based on the decomposition theorem, all plant species were divided, under relation $\hat{R}_1$, into clusters of the similar species based on their state. Results of fuzzy clusterization under similarity relation $\hat{R}_1$, are provided in Table 3.18.

The results obtained allow concluding that the state of "European white birch" plants essentially differs from the states of all other plant species. States of "Witch elm" and "large-leaved linden" plants, and also states of "little-leaved linden" and "Hungarian lilac" plants are similar to each other. Based on the states of other plant species one can say with high confidence level that there is no similarity of them.

Table 3.18 Clusterization of plant species under similarity relation $\hat{R}_1$

Confidence level	Cluster
0.495	{1}{2,3,4,5,6,7,8,9,10,11,12,13,14,15,16,17}
0.766	{1}{2,9,11,12,17}{3,4,5,6,7,8,10,13,14,15,16}
0.818	{1}{2,9,11,12,17}{3,5,6}{4,7,8,10,13,14,15,16}
0.820	{1}{2,9,11,12,17}{3,5}{6}{4,7,8,10,13,14,15,16}
0.825	{1}{2}{3,5}{6}{9,11,12,17}{4,7,8,10,13,14,15,16}
0.831	{1}{2}{3,5}{6}{8}{9,11,12,17}{4,7,10,13,14,15,16}
0.844	{1}{2}{3,5}{6}{8}{9,11,12,17}{7,15}{14}{4,10,13,16}
0.869	{1}{2}{3,5}{6}{8}{9}{4,10,13}{11,12,17}{7,15}{14}{16}
0.872	{1}{2}{3,5}{6}{8}{9}{4,10,13}{11,12}{7,15}{14}{16}{17}
0.878	{1}{2}{3,5}{6}{7}{8}{9}{4,10}{11,12}{13}{14}{15}{16}{17}
0.901	{1}{2}{3,5}{4}{6}{7}{8}{9}{10}{11,12}{13}{14}{15}{16}{17}
0.918	{1}{2}{3}{4}{5}{6}{7}{8}{9}{10}{11,12}{13}{14}{15}{16}{17}
0.964	{1}{2}{3}{4}{5}{6}{7}{8}{9}{10}{11}{12}{13}{14}{15}{16}{17}

Chapter 4
Model-Building Techniques of the Generalized Characteristic Expert Evaluation Models

4.1 The Generalized Model of Characteristic Expert Evaluations Based on the Least Square Method

The expert evaluations theory has formulated the optimum condition of group sampling as per Pareto [135]. This condition means that if $R = F(R_1, R_2,..., R_k)$ is the group ranking, which is function of individual rankings $R_1, R_2,..., R_k$, then

$$\bigcap_{n=1}^{k} R_n \subseteq R \subseteq \bigcup_{n=1}^{k} R_n.$$

Let $\Xi^k = \{X_i; i = \overline{1,k}\}$, where X_i; $i = \overline{1,k}$ be models of expert evaluations of qualitative characteristic or expert description in linguistic terms of physical values of quantitative characteristic with membership functions of term-sets $\{\mu_{il}(x), l = \overline{1,m}\}$, $\mu_{il}(x) \equiv (a_1^{il}, a_2^{il}, a_L^{il}, a_R^{il})$. Let us construct generalized model of X expert evaluations of a property based on the models X_i [COSS with membership functions of term-set $f_l(x)$, $f_l(x) = (a_1^l, a_2^l, a_L^l, a_R^l)$].

Let us formulate Pareto condition for the optimum generalized model of expert evaluations of a qualitative characteristic or expert description of values of quantitative characteristic in the linguistic terms, which is constructed on the basis of models X_i

$$\bigcap_{i=1}^{k} X_i \subseteq X \subseteq \bigcup_{i=1}^{k} X_i.$$

or

$$\min_{i=1,k}[\mu_{1l}(x), \mu_{2l}(x),..., \mu_{kl}(x)] \leq f_l(x) \leq$$

$$\leq \max_{i=1,k}[\mu_{1l}(x), \mu_{2l}(x),..., \mu_{kl}(x)]; \quad \forall x \in [0,1], \; l = \overline{1,m}. \qquad (4.1)$$

Let us assume [150] that membership functions of a term-set of the generalized characteristic expert evaluations model is related to the same class of functions as

O.M. Poleshchuk and E.G. Komarov: Expert Fuzzy Info. Processing, STUDFUZZ 268, pp. 119–145.
springerlink.com   © Springer-Verlag Berlin Heidelberg 2011

membership functions of the term-sets of the set Ξ^k elements; i.e. if the fuzzy numbers corresponding to the term-sets of the set Ξ^k elements are tolerance or unimodal numbers of the group Λ described in §2.1, then fuzzy numbers corresponding to a term-set of the generalized model are also identified as tolerance or unimodal numbers of the group Λ. Let us suppose that weight coefficients ω_i of of set Ξ^k elements are revealed. Let us define parameters of membership functions $(a_1^l, a_2^l, a_L^l, a_R^l)$ of a term-set of the generalized model of expert qualitative characteristic evaluations or expert description of quantitative characteristic values in linguistic terms, from the condition

$$F = \sum_{l=1}^{m} \sum_{i=1}^{k} \omega_i \left[\left(a_1^{il} - a_1^l \right)^2 + \left(a_2^{il} - a_2^l \right)^2 + \left(a_L^{il} - a_L^l \right)^2 + \left(a_R^{il} - a_R^l \right)^2 \right] \to \min. \quad (4.2)$$

Unknown parameters are determined from system of normal equations at $l = \overline{1, m}$

$$\frac{\partial F}{\partial a_1^l} = 2 \left[\sum_{i=1}^{k} \omega_i a_1^{il} - a_1^l \right] = 0; \quad \frac{\partial F}{\partial a_2^l} = 2 \left[\sum_{i=1}^{k} \omega_i a_2^{il} - a_2^l \right] = 0;$$

$$\frac{\partial F}{\partial a_L^l} = 2 \left[\sum_{i=1}^{k} \omega_i a_L^{il} - a_L^l \right] = 0; \quad \frac{\partial F}{\partial a_R^l} = 2 \left[\sum_{i=1}^{k} \omega_i a_R^{il} - a_R^l \right] = 0. \quad (4.3)$$

We obtain solutions

$$a_1^l = \sum_{i=1}^{k} \omega_i a_1^{il}; \quad a_2^l = \sum_{i=1}^{k} \omega_i a_2^{il}; \quad a_L^l = \sum_{i=1}^{k} \omega_i a_L^{il}; \quad a_R^l = \sum_{i=1}^{k} \omega_i a_R^{il}. \quad (4.4)$$

According to optimization problem solutions, the generalized model is a linear combination of the set Ξ^k elements (see §3.1). Coefficients of the linear combination are weight coefficients of the characteristic expert evaluations models which are used to construct the generalized model. Thus, we obtain

$$X = \{ f_l(x) \} = \left\{ f_l(x) = \sum_{i=1}^{k} \omega_i \mu_{il}(x) \right\} =$$

$$= \left\{ f_l(x) \equiv \left(\sum_{i=1}^{k} \omega_i a_1^{il}, \sum_{i=1}^{k} \omega_i a_2^{il}, \sum_{i=1}^{k} \omega_i a_L^{il}, \sum_{i=1}^{k} \omega_i a_R^{il} \right) \right\}.$$

$$(4.5)$$

Let us denote the constructed generalized model of expert qualitative characteristic evaluations or expert description, in linguistic terms, of physical values of quantitative characteristic as optimum by noisiness of the set Ξ^k element parameters.

Let us prove the fulfillment of the condition (4.1) ensuring Pareto optimality of generalized model of an expert evaluation or description of a characteristic. As

$$\min_{i=1,k}\left[\mu_{1l}(x),\mu_{2l}(x),...,\mu_{kl}(x)\right]=\sum_{i=1}^{k}\omega_i\left(\min_{i=1,k}\left[\mu_{1l}(x),\mu_{2l}(x),...,\mu_{kl}(x)\right]\right)\le$$

$$\le f_l(x)=\sum_{i=1}^{k}\omega_i\mu_{il}(x)\le\sum_{i=1}^{k}\omega_i\left(\max_{i=1,k}\left[\mu_{il}(x),\mu_{2l}(x),...,\mu_{kl}(x)\right]\right)=$$

$$=\max_{i=1,k}\left[\mu_{1l}(x),\mu_{2l}(x),...,\mu_{kl}(x)\right];\ \forall x\in[0,1],\ l=\overline{1,m},$$

$$(4.6)$$

then from (4.6) it is obtained

$$\min_{i=1,k}\left[\mu_{1l}(x),\mu_{2l}(x),...,\mu_{kl}(x)\right]\le f_l(x)\le$$

$$\le\max_{i=1,k}\left[\mu_{1l}(x),\mu_{2l}(x),...,\mu_{kl}(x)\right];\ \forall x\in[0,1],\ l=\overline{1,m}.$$

Thus, the generalized model of expert qualitative characteristic evaluations or expert description, in linguistic terms, of physical values of the quantitative characteristic constructed in §4.1 with use of the least squares method, is Pareto optimal one.

4.2   Definition of Weight Coefficients of the Formalized Results of Expert Qualitative Characteristic Evaluations Based on the Similarity Relations

Let similarity relations with matrixes $R_1=\left[\mu_{R_1}(X_i,X_j),i=\overline{1,k},j=\overline{1,k}\right]$; $R_1=\left[\mu_{R_1}(M_i,M_j),i=\overline{1,k},j=\overline{1,k}\right]$, accordingly, and conformity relations with matrixes $\hat{R}_1=\mu_{R_1}(X_i,X_j)$; $\hat{R}_1=\mu_{R_1}(M_i,M_j)$, accordingly, are constructed on sets Ξ^k and Θ^k using similarity indexes.

Let similarity relations with matrixes $R_2=\left[\mu_{R_2}(X_i,X_j)\right]$; $R_2=\left[\mu_{R_2}(M_i,M_j)\right]$, accordingly, and conformity relation with matrix $\hat{R}_2=\mu_{R_2}(X_i,X_j)$; $\hat{R}_2=\mu_{R_2}(M_i,M_j)$ are constructed based on sets Ξ^k and Θ^k using consistency indexes. According to the decomposition theorem, the matrix $\hat{R}_p$, $p=\overline{1,2}$ can de decomposed onto equivalence relations

$$\hat{R}_p=\max_{\alpha}\left\{\alpha\begin{pmatrix}1 & \delta_{12} & \cdots & \delta_{1k}\\ \delta_{12} & 1 & \cdots & \delta_{2k}\\ \cdot & \cdot & \cdots & \cdot\\ \delta_{1k} & \delta_{2k} & \cdots & 1\end{pmatrix}\right\},p=\overline{1,2},$$

where $\delta_{ij}=\begin{cases}0;\\1,\end{cases}\ i=\overline{1,k},\ j=\overline{1,k}.$

Thus, depending on confidence level α, set Ξ^k (COSS set) or Θ^k set (set of formalized expert evaluation results of a qualitative characteristic of an object group) can be divided into clusters of similar elements.

Let us define weight coefficients of elements of sets Ξ^k and Θ^k depending on confidence level within the limits of following criteria:

I.   Weight coefficient of the element which has entered a larger cluster exceeds weight coefficient of the element which has entered a smaller cluster.
II. If elements have entered the same cluster, their weight coefficients are equal.

Let us consider various cases:

1. With confidence level $\alpha < 1$ all elements enter the same cluster.
   Based on the criterion II, their weight coefficients

$$\omega_i, i = \overline{1,k}, \sum_{i=1}^{k} \omega_i = 1$$

are considered equal, i.e.

$$\omega_i = 1/k, i = \overline{1,k}.$$

2. With confidence level $\alpha < 1$ two clusters of potency (number of elements): $k-1$ and 1 occur.

Let us assume that elements with indexes $i = \overline{1,k-1}$ have entered a cluster of potency $k-1$. Based on the criterion II, weight indexes ω_i are considered equal, and based on the criterion I, the weight coefficient of the element with the index k has smaller value than values of weight coefficients

$$\omega_i, \quad \sum_{i=1}^{k} \omega_i = 1.$$

Let us use Fishburn scale [151] according to which weight coefficients of units ranked in decrease of their importance order (within the limits of certain criterion) are determined under the formula:

$$\omega_i = \frac{2(k-i+1)}{k(k+1)}, i = \overline{1,k}. \tag{4.7}$$

Since the weight coefficient of an element with k index is the least one, it is determined by substitution $i = k$ in the formula (4.7), hence we obtain

$$\omega_k = \frac{2}{k(k+1)}. \tag{4.8}$$

Let us compute the sum of weight coefficients of elements with indexes $i = \overline{1,k-1}$ using the formula (4.7):

$$\sum_{i=1}^{k-1} \omega_i = \sum_{i=1}^{k-1} \frac{2(k-i+1)}{k(k+1)} = \left[\frac{2k+4}{2k(k+1)}\right](k-1) = \left[\frac{k+2}{k(k+1)}\right](k-1). \tag{4.9}$$

As elements with these indexes have equal weight coefficients, we obtain

$$\omega_i = \frac{k+2}{k(k+1)}, i = \overline{1, k-1}. \tag{4.10}$$

3. With confidence level $\alpha < 1$ there are several clusters: one cluster of potency v and other clusters of potency 1.

The largest cluster makes a basis, and according to the criterion II the elements belonging to this cluster are considered to have equal weight coefficients. According to the criterion I, elements of individual clusters have weight coefficients less than weight coefficients of elements of the largest cluster. Let us assume that elements with indexes $i = \overline{1, v}$ have entered the largest cluster. Let us rank the elements which have entered individual clusters by values of indexes ρ_j or $\breve{\rho}_j$ with $j = \overline{v+1, k}$

$$\rho_j = \sum_{i=1}^{v} \mu_{R_1}\left(X_i, X_j\right)\left[\rho_j = \sum_{i=1}^{v} \mu_{R_1}\left(M_i, M_j\right)\right],$$

or

$$\breve{\rho}_j = \sum_{i=1}^{v} \mu_{R_2}\left(X_i, X_j\right)\left[\breve{\rho}_j = \sum_{i=1}^{v} \mu_{R_2}\left(M_i, M_j\right)\right]. \tag{4.11}$$

Indexes ρ_j or $\breve{\rho}_j$ of (4.11) are summarized quantity indexes of similarity (consistency) of elements $X_i(M_i)$, $i = \overline{v+1, k}$, entering individual clusters, with elements from a cluster with potency v. Let us consider that the more ρ_j $\left(\breve{\rho}_j\right)$ is, the more importance of element $X_j(M_j)$ is and, accordingly, the more its weight coefficient is. Let us obtain a conditional ordered series of elements Ξ^k ranged by lack of growth of their weight coefficients

$$X_1 = X_2 = \ldots = X_v > X_{(v+1)} > X_{(v+2)} > \ldots > X_{(k)}$$

or a conditional ordered series of elements Θ^k, ranged in the same manner

$$M_1 = M_2 = \ldots = M_v > M_{(v+1)} > M_{(v+2)} > \ldots > M_{(k)}.$$

Weight coefficients of elements with indexes $i = \overline{(v+1), k}$ are determined by consecutive substitution of indexes $i = \overline{v+1, k}$ into the formula (4.7). The result is

$$\omega_{(j)} = \frac{2(k - j + 1)}{k(k+1)}, j = \overline{v+1, k}. \tag{4.12}$$

Let us compute the sum of weight coefficients of elements with indexes $i = \overline{1,v}$ from the formula (4.7):

$$\sum_{i=1}^{v}\omega_i = \sum_{i=1}^{v}\frac{2(k-i+1)}{k(k+1)} = v\cdot\left(\frac{2k+2k-2v+2}{2k(k+1)}\right) = v\cdot\left(\frac{2k-v+1}{k(k+1)}\right). \quad (4.13)$$

Since elements with these indexes have equal weight coefficients, that on dividing (4.13) by v, we obtain

$$\omega_i = \frac{2k-v+1}{k(k+1)}. \quad (4.14)$$

4. With confidence level $\alpha < 1$ some clusters occur, one of them is large cluster with potency v and several clusters which potency is less than v; with that all small clusters have different potencies.

Let the largest of small clusters have potency d, and other clusters, without limiting their generality, have potencies $d > a > b > c \geq 1$, accordingly. Applying the criteria I and II, we range elements according to their weight coefficients. We obtain a conditional ordered series of elements Ξ^k (or elements Θ^k)

$$X_{(1)} = X_{(2)} = ... = X_{(v)} > X_{(v+1)} = ... = X_{(v+d)} > X_{(v+d+1)} = ... = X_{(v+d+a)} >$$
$$> X_{(v+d+a+1)} = ... = X_{(v+d+a+b)} > X_{(v+d+a+b+1)} = ... = X_{(k)}.$$

Weight coefficients of elements of each cluster are computed in the same manner as the weight coefficients of elements of a cluster with potency v (see item 3). We obtain

$$\omega_{(i)} = \frac{2k-v+1}{k(k+1)}, i = \overline{1,v}; \quad \omega_{(i)} = \frac{2k-2v-d+1}{k(k+1)}, i = \overline{v+1,v+d};$$

$$\omega_{(i)} = \frac{2k-2v-2d-a+1}{k(k+1)}, i = \overline{v+d+1,v+d+a};$$

$$\omega_{(i)} = \frac{2k-2v-2d-2a-b+1}{k(k+1)}, i = \overline{v+d+a+1,v+d+a+b};$$

$$\omega_{(i)} = \frac{2k-2v-2d-2a-2b-c+1}{k(k+1)}, i = \overline{v+d+a+b+1,k}.$$

5. With confidence level $\alpha < 1$ some large clusters with identical amount of elements occur:
   a) Two clusters of identical potency v occur. Let us assume that the first cluster consists of elements with indexes $i = \overline{1,v}$, and the second cluster consists of elements with indexes $i = \overline{v+1,2v}$. Other elements are fractionalized into individual clusters. Let us calculate indexes ρ_j or $\breve{\rho}_j$ for elements of two clusters of potency v with $j = \overline{1,2v}$

$$\rho_j = \sum_{i=1}^{2v} \mu_{R_1}\left(X_i, X_j\right)\left[\rho_j = \sum_{i=1}^{2v} \mu_{R_1}\left(M_i, M_j\right)\right];$$

$$\breve{\rho}_j = \sum_{i=1}^{2v} \mu_{R_2}\left(X_i, X_j\right)\left[\breve{\rho}_j = \sum_{i=1}^{2v} \mu_{R_2}\left(M_i, M_j\right)\right]. \tag{4.15}$$

Let us select the maximum index from indexes ρ_j, $j = \overline{1,2v}$ $\left(\breve{\rho}_j, j = \overline{1,2v}\right)$ and, considering that an element corresponding to this index belongs to the first cluster of potency v, we obtain a conditional ordered series of elements Ξ^k (or elements Θ^k)

$$X_1 = X_2 = \ldots = X_v > X_{v+1} = X_{v+2} = \ldots = X_{2v} > X_{(2v+1)} > \ldots > X_{(k)}.$$

Individual clusters are ranged similarly to ranging of individual clusters in p. 3. In this case it is easy to demonstrate that

$$\omega_i = \frac{2k - v + 1}{k(k+1)}, i = \overline{1,v}; \quad \omega_i = \frac{2k - 3v + 1}{k(k+1)}, i = \overline{v+1, 2v};$$

$$\omega_{(i)} = \frac{2(k - i + 1)}{k(k+1)}, i = \overline{2v+1, k}; \tag{4.16}$$

b) A set of elements with indexes $i = \overline{2v+1, k}$ is fractionalized into clusters of different potency. The potency of each cluster is less than v, but it is more than '1' or equal to '1'. In this case the most powerful cluster is selected, and the procedure from p. 4 is carried out to determine weight coefficients of elements with indexes $i = \overline{2v+1, k}$. Weight coefficients of elements of clusters of potency v remain unchanged

$$\omega_i = \frac{2k - v + 1}{k(k+1)}, i = \overline{1,v}; \quad \omega_i = \frac{2k - 3v + 1}{k(k+1)}, i = \overline{v+1, 2v}; \tag{4.17}$$

c) The set of elements with indexes $i = \overline{2v+1, k}$ is fractionalized into clusters with potencies less than v each, but more or equal to '1'y, and among these clusters there are several large ones identical in their potencies. Let us consider that there are two clusters of potency $d > 1$ with elements $X_j(M_j)$, $j = \overline{2v+1, 2v+2d}$. For these elements indexes ρ_j or $\breve{\rho}_j$ are defined.

$$\rho_j = \sum_{i=1}^{v} \mu_{R_1}\left(X_i, X_j\right)\left[\rho_j = \sum_{i=1}^{v} \mu_{R_1}\left(M_i, M_j\right)\right];$$

$$\breve{\rho}_j = \sum_{i=1}^{v} \mu_{R_2}\left(X_i, X_j\right)\left(\breve{\rho}_j = \sum_{i=1}^{v} \mu_{R_2}\left(M_i, M_j\right)\right). \tag{4.18}$$

Then, the procedure of the weight coefficients computation described in p. 5a is carried out. Weight coefficients of elements from clusters of potency v remain unchanged

$$\omega_i = \frac{2k - v + 1}{k(k+1)}, i = \overline{1, v}; \quad \omega_i = \frac{2k - 3v + 1}{k(k+1)}, i = \overline{v, 2v};$$

(4.19)

6. With the confidence level equal to '1', k clusters occur whose elements are ranged according to values ρ_j, or $\breve{\rho}_j$, $j = \overline{1, k}$

$$\rho_j = \sum_{i=1}^{k} \mu_{R_1}\left(X_i, X_j\right)\left(\rho_j = \sum_{i=1}^{k} \mu_{R_1}\left(M_i, M_j\right)\right);$$

$$\breve{\rho}_j = \sum_{i=1}^{k} \mu_{R_2}\left(X_i, X_j\right)\left(\breve{\rho}_j = \sum_{i=1}^{k} \mu_{R_2}\left(M_i, M_j\right)\right).$$

Let us range values ρ_j, $j = \overline{1, k}$ $\left(\breve{\rho}_j, j = \overline{1, k}\right)$ in decreasing order. We obtain a conditional ordered series

$$X_{(1)} > X_{(2)} > ... > X_{(i)} > ... > X_{(k)}$$

or

$$M_{(1)} > M_{(2)} > ... > M_{(i)} > ... > M_{(k)}.$$

Weight coefficients of sets Ξ^k and Θ^k elements are computed by the formula (4.7).

If any $\rho_j(\breve{\rho}_j)$ are equal, then we obtain a conditional equality $X_{(1)} = ... = X_{(l)}$. In this case

$$\omega_{(j)} = \frac{1}{l}\sum_{i=1}^{l}\frac{2(k - j + 1)}{k(k+1)}, \quad j = \overline{1, l}.$$

(4.20)

4.3 Determination of Weight Coefficients for Models of Characteristic Expert Evaluations Based on the Fuzziness Degrees

The selection of a principle for determination of weight coefficients depends on particular situation, requirements etc. For example, determination of weight coefficients for models of expert qualitative characteristic evaluations or expert description, in linguistic terms, of physical values of quantitative characteristic (elements of set Ξ^k) can be carried out on the basis of degrees of the models fuzziness (COSS). Fuzziness degree of an evaluation model or characteristic description, as known [28], is a quantity index of average degree of difficulties an expert undergoes while describing and estimating real objects within the scope of relevant set of linguistic values of the characteristic considered. Therefore we believe logical to assign the greatest weight coefficient to an element Ξ^k (i.e. to the model of expert evaluations of characteristic) with the minimum fuzziness degree, and the least weight coefficient to an element Ξ^k with maximum fuzziness degree. If fuzziness degrees of those elements are equal, weight coefficients of those elements are considered equal.

Fuzziness degree of a characteristic expert evaluations model constructed in the form of COSS is defined as follows:

$$\zeta = \frac{1}{|U|} \int_U f\left[\mu_{k_1}(x) - \mu_{k_2}(x)\right] dx,$$

where U is a universal set (a subset on the real number line)

$$\mu_{k_1}(x) = \max_{1 \leq l \leq m} \mu_l(x); \quad \mu_{k_2}(x) = \max_{\substack{1 \leq l \leq m \\ k \neq k_1}} \mu_l(x);$$

f decreases, and $f(0) = 1$, $f(1) = 0$.

Let us be limited to reviewing linear membership functions on a set

$$\overline{U} = \bigcup_l \left\{ x \in U : 0 < \mu_l(x) < 1, l = \overline{1, m} \right\},$$

and let us assume an integrand be equal $f(x) = 1 - x$. It is the unique linear function satisfying to conditions: f decreases, $f(0) = 1$, $f(1) = 0$. Then according to [28]

$$\zeta = |\overline{U}| / 2|U|.$$

Let us calculate fuzziness degree for each element of set Ξ^k and range all its elements as per the following principle: the less fuzziness degree of an element is, the higher its rank is. Without limiting its generality, we obtain, for example, a conditional ordered series

$$X_{(1)} > X_{(1)} > \ldots > X_{(i)} > \ldots > X_{(k)}.$$

Weight coefficients are computed by the formula (4.7).

If any elements have identical fuzziness degrees, determination weight coefficients is made within the limits of schemes of §4.2.

If fuzziness degrees of all elements are equal, these elements are considered equivalent, and weight coefficients

$$\omega_i, i = \overline{1, k}; \quad \sum_{i=1}^{k} \omega_i = 1,$$

equal, i.e

$$\omega_i = 1 / k, i = \overline{1, k}.$$

4.4 Building of the Generalized Model of Characteristic Expert Evaluations Based on the Minimum Information Loss

Let $X_i = \{\mu_{il}(x), l = \overline{1,m}\}$, $i = \overline{1,k}$ be elements of set Ξ^k (models of expert evaluations of qualitative characteristic or expert description of quantitative characteristic physical values, in linguistic terms), $\mu_{il}(x) \equiv (a_1^{il}, a_2^{il}, a_L^{il}, a_R^{il})$.

Let us denote the generalized model within the limits of elements of set Ξ^k defined as COSS $X = \{f_l(x), l = \overline{1,m}\}$, $f_l(x) = (a_1^l, a_2^l, a_L^l, a_R^l)$, $l = \overline{1,m}$.

In §3.1 the distinction index of two elements X_i and X_j of set Ξ^k with membership functions $\{\mu_{il}(x), l = \overline{1,m}\}$, $\{\mu_{jl}(x), l = \overline{1,m}\}$, $i = \overline{1,k}$, $j = \overline{1,k}$, accordingly, or information loss between elements X_i and X_j of set Ξ^k

$$d(X_i, X_j) = \frac{1}{2} \sum_{l=1}^{m} \int_0^1 |\mu_{il}(x) - \mu_{jl}(x)| dx$$

is defined.

By analogy to this definition, let us introduce definition of information loss between generalized model $X = \{f_l(x), l = \overline{1,m}\}$ and an element $X_i = \{\mu_{il}(x), l = \overline{1,m}\}$, $i = \overline{1,k}$ of set Ξ^k, i.e.

$$d(X_i, X) = \frac{1}{2} \sum_{l=1}^{m} \int_0^1 |\mu_{il}(x) - f_l(x)| dx.$$

Let us denote average value of information losses between elements of set Ξ^k and the generalized model

$$\sigma = \frac{1}{k} \sum_{i=1}^{k} d(X_i, X), i = \overline{1,k}.$$

As information loss occurred while constructing the generalized model within the scope of the set Ξ^k.

Let us consider that fuzzy numbers with membership functions $\mu_{il}(x) \equiv (a_1^{il}, a_2^{il}, a_L^{il}, a_R^{il})$, $i = \overline{1,k}$, $f_l(x) = (a_1^l, a_2^l, a_L^l, a_R^l)$, $l = \overline{1,m}$, corresponding to term-sets of set Ξ^k elements and to term-set of the generalized model, are T-numbers or normal triangular numbers, $U = [0,1]$.

Let us introduce new parameters of membership functions of term-sets of set Ξ^k elements and membership functions of term-set of the generalized model, which are abscissas of breakpoints of graphs of these membership functions [150]:

$$a_{il1} = a_1^{il} - a_L^{il}, \; a_{il2} = a_1^{il}, \; a_{il3} = a_2^{il}, \; a_{il4} = a_2^{il} + a_R^{il},$$

$$a_{l1} = a_1^l - a_L^l, \; a_{l2} = a_1^l, \; a_{l3} = a_2^l, \; a_{l4} = a_2^l + a_R^l.$$

As $a_{i11} = 0$, $a_{i12} = 0$, $i = \overline{1,k}$, we suppose that $a_{11} = 0$, $a_{12} = 0$. As $a_{im3} = 1$, $a_{im4} = 1$, $i = \overline{1,k}$, we suppose that $a_{m3} = 1$, $a_{m4} = 1$.

Let us denote a half-sum of a module integral of difference between related right boundaries of membership functions of a set Ξ^k element l-th term and the generalized model, on the one hand, and an module integral of difference between related left boundaries of membership functions of a set Ξ^k element $(l+1)$-th term and the generalized model $l = \overline{1, m-1}$, on the other hand, as information loss within boundaries of l-th and $(l+1)$-th terms between a set Ξ^k element and its generalized model.

Let us consider various cases of disposition of boundaries of membership functions (Fig. 4.1) of the adjacent terms of set Ξ^k i-th element and boundaries of membership functions of the same terms of the generalized model. Let us determine information losses depending on the disposition of boundaries of membership functions.

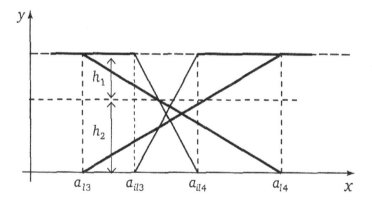

Fig. 4.1 Boundaries of membership functions

If $a_{l3} > a_{il3}$, $a_{l4} > a_{il4}$, then loss of the information within the boundaries of l-th and $(l+1)$-th terms is equal to square of trapezoid with base $a_{l3} - a_{il3}$, $a_{l4} - a_{il4}$ and unit height, i.e.

$$\frac{1}{2}\left(a_{l3} - a_{il3} + a_{l4} - a_{il4}\right). \tag{4.21}$$

If $a_{l3} < a_{il3}$, $a_{l4} < a_{il4}$, information loss on boundary of $(l+1)$-th terms is equal to square of trapezoid with base $a_{il3} - a_{l3}$, $a_{il4} - a_{l4}$ and unit height, i.e.

$$\frac{1}{2}\left(-a_{l3} + a_{il3} - a_{l4} + a_{il4}\right). \tag{4.22}$$

If $a_{l3} \leq a_{il3}$, $a_{l4} \geq a_{il4}$, then loss of the information within the boundaries of l-st and $(l+1)$-th terms is equal to the sum of squares of two triangles.

One triangle has its base equal to $a_{il3} - a_{l3}$, and another triangle has its base equal to $a_{l4} - a_{il4}$, too. Let us determine heights of these triangles.

As triangles with bases $a_{il3} - a_{l3}$ and $a_{l4} - a_{il4}$ are similar, we have

$$\begin{cases} \dfrac{h_1}{h_2} = \dfrac{a_{il3} - a_{l3}}{a_{l4} - a_{il4}}; \\ h_1 + h_1 = 1, \end{cases}$$

where h_1 and h_2 are heights of corresponding triangles. Hence, the height of triangles with the base $a_{il3} - a_{l3}$ is equal to

$$h_1 = \frac{a_{il3}a_{l4} - a_{il3}a_{l3} - a_{l3}a_{il4} + a_{l3}a_{il3} - a_{l3}a_{l4} + a_{l3}^2 + a_{l3}a_{il4} - a_{l3}a_{il3}}{(a_{l4} - a_{l3})(a_{l4} - a_{l3} - a_{il4} + a_{il3})} =$$

$$= \frac{a_{il3} - a_{l3}}{a_{l4} - a_{l3} - a_{il4} + a_{il3}}.$$

Height of triangles with the base $a_{l4} - a_{il4}$ is equal to

$$h_1 = 1 - \frac{a_{il3} - a_{l3}}{a_{l4} - a_{l3} - a_{il4} + a_{il3}} = \frac{a_{l4} - a_{il4}}{a_{l4} - a_{l3} - a_{il4} + a_{il3}}.$$

In this case information loss is equal to

$$\frac{(a_{l3} - a_{il3})^2 + (a_{l4} - a_{il4})^2}{2(a_{l4} - a_{il4} + a_{il3} - a_{l3})}. \tag{4.23}$$

If $a_{l3} > a_{il3}$, $a_{l4} < a_{il4}$, then information loss is equal to

$$-\frac{(a_{l3} - a_{il3})^2 + (a_{l4} - a_{il4})^2}{2(a_{l4} - a_{il4} + a_{il3} - a_{l3})}. \tag{4.24}$$

Thus, from (4.20) — (4.24) it follows that the general loss of the information

$$\sigma = \frac{1}{k} \sum_{l=1}^{m-1} \sum_{i=1}^{k} \left\{ \frac{1}{2} \delta_{il}^1 (a_{l3} - a_{il3} + a_{l4} - a_{il4}) + \delta_{il}^2 \left[\frac{(a_{l3} - a_{il3})^2 + (a_{l4} - a_{il4})^2}{2(a_{l4} - a_{il4} + a_{il3} - a_{l3})} \right] \right\}, \tag{4.25}$$

$$\delta_{il}^{1} = \begin{cases} 1, & a_{l3} \geq a_{il3}, a_{l4} \geq a_{il4}; \\ -1, & a_{l3} \leq a_{il3}, a_{l4} \leq a_{il4}; \\ 0, & a_{l3} > a_{il3}, a_{l4} < a_{il4} \text{ or } a_{l3} < a_{il3}, a_{l4} > a_{il4}; \end{cases}$$

$$\delta_{il}^{2} = \begin{cases} 1, & a_{l3} < a_{il3}, a_{l4} > a_{il4}; \\ -1, & a_{l3} > a_{il3}, a_{l4} < a_{il4}; \\ 0, & a_{l3} \geq a_{il3}, a_{l4} \geq a_{il4} \text{ or } a_{l3} \leq a_{il3}, a_{l4} \leq a_{il4}; \end{cases}$$

Unknown parameters a_{l3}, a_{l4}, $l = \overline{1, m-1}$ are solutions of optimization problem [150]

$$\sigma = \frac{1}{2k} \sum_{l=1}^{m-1} \sum_{i=1}^{k} \left[\delta_{il}^{1}(a_{l3} - a_{il3} + a_{l4} - a_{il4}) + \delta_{il}^{2}\left(\frac{(a_{l3} - a_{il3})^2 + (a_{l4} - a_{il4})^2}{a_{l4} - a_{il4} + a_{il3} - a_{l3}} \right) \right] \rightarrow \min.$$

Solutions meet limits of known methods (152].

The generalized model of expert evaluations of qualitative characteristic or expert description of quantitative property values in linguistic terms, being constructed within the scope of set Ξ^k elements, maintains a maxima of the information included in the set elements. However, unlike the model constructed in §4.2, it generally does not satisfy to Pareto condition. In particular practical problems, the fulfillment of Pareto condition for the generalized model constructed on the basis of a information loss minimum, is directly verified. Besides, the method of determination of generalized model described in §4.2 can be applied within the limits of any membership function of fuzzy numbers from Λ group, and the method of the current paragraph can be applied only in case of T-numbers or normal triangular numbers.

Thus, for practical problems, while determining the generalized models within the limits of set Ξ^k elements, we propose to define models according to methods described in §4.2 and §4.4. Then, it is necessary to check Pareto optimality of the generalized model constructed on the basis of information loss minimum. After that, on the basis of two characteristics, namely, fuzziness degrees of models and information loss occurred while constructing the models, we can define the optimum generalized model.

4.5   Building of the Generalized Formalized Result of Expert Evaluations of the Qualitative Characteristic Based on the Least Squares Method

Let $M_i = \{\mu_i^n, n = \overline{1, N}\}$, $i = \overline{1, k}$ be elements of set Θ^k (the formalized results of expert evaluations of qualitative characteristic of an object group), $\mu_i^n \equiv \left(a_1^{in}, a_2^{in}, a_L^{in}, a_R^{in} \right)$, $n = \overline{1, N}$. Let us construct the generalized formalized outcome of the expert evaluation of qualitative characteristic of N objects group $M = \{\mu^n, n = \overline{1,N}\}$, $\mu^n(x) = (a_1^n, a_2^n, a_L^n, a_R^n)$, $n = \overline{1,N}$ on the basis of the formalized results M_i of an evaluation made by k experts of an object group within the scope of a qualitative characteristic.

Let us formulate Pareto condition for the optimum generalized result of M expert evaluation of qualitative characteristic of an object group constructed on the basis of elements M_i, $i = \overline{1,k}$ of the set Θ^k:

$$\bigcap_{i=1}^{k} M_i \subseteq M \subseteq \bigcup_{i=1}^{k} M_i$$

or

$$\min_{i=1,k}\left[\mu_1^n(x), \mu_2^n(x), \dots, \mu_k^n(x)\right] \leq \mu^n(x) \leq$$
$$\leq \max_{i=1,k}\left[\mu_1^n(x), \mu_2^n(x), \dots, \mu_k^n(x)\right], \ \forall x \in [0,1], \ n = \overline{1,N}. \tag{4.26}$$

Let us assume that membership functions of the generalized formalized result of an expert evaluation of qualitative characteristic of units group $M = \{\mu^n, n = \overline{1,N}\}$ belong to the same class of functions as membership functions of set Θ^k elements, i.e. if μ_i^n are membership functions of tolerance or unimodal numbers from an Λ group, then μ^n, $n = \overline{1,N}$ are also defined as membership functions of tolerance or unimodal numbers from the Λ group.

Let within the limits of a method 4.3 weight coefficients ω_i, $i = \overline{1,k}$ of the set Θ^k elements are defined. We determine parameters $a_1^n, a_2^n, a_L^n, a_R^n$, $i = \overline{1,N}$ of membership functions of the generalized formalized outcome of an expert evaluation of qualitative characteristic of an object group from the condition

$$\Phi = \sum_{n=1}^{N} \sum_{i=1}^{k} \omega_i\left[\left(a_1^{in} - a_1^n\right)^2 + \left(a_2^{in} - a_2^n\right)^2 + \left(a_L^{in} - a_L^n\right)^2 + \left(a_R^{in} - a_R^n\right)^2\right] \to \min. \tag{4.27}$$

Parameters are determined from the system of normal equations with $n = \overline{1,N}$:

$$\frac{\partial F}{\partial a_1^n} = 2\left[\sum_{i=1}^{k} \omega_i a_1^{in} - a_1^n\right] = 0; \quad \frac{\partial F}{\partial a_2^n} = 2\left[\sum_{i=1}^{k} \omega_i a_2^{in} - a_2^n\right] = 0;$$

$$\frac{\partial F}{\partial a_L^n} = 2\left[\sum_{i=1}^{k} \omega_i a_L^{in} - a_L^n\right] = 0; \quad \frac{\partial F}{\partial a_R^n} = 2\left[\sum_{i=1}^{k} \omega_i a_R^{in} - a_R^n\right] = 0. \tag{4.28}$$

We obtain the solutions:

$$a_1^n = \sum_{i=1}^{k} \omega_i a_1^{in}; \quad a_2^n = \sum_{i=1}^{k} \omega_i a_2^{in}; \quad a_L^n = \sum_{i=1}^{k} \omega_i a_L^{in}; \quad a_R^n = \sum_{i=1}^{k} \omega_i a_R^{in}. \tag{4.29}$$

Thus, the generalized formalized result of an expert evaluation of qualitative characteristic of an object group constructed within the limits of set Θ^k elements is a linear combination of these elements. Linear combination coefficients are weight coefficients of set Θ^k elements.

Thus,

$$M = \left\{\mu^n(x), n = \overline{1,N}\right\} = \left\{\sum_{i=1}^{k} \omega_i \mu_i^n(x), n = \overline{1,N}\right\} =$$

$$= \left\{\mu^n(x) \equiv \left(\sum_{i=1}^{k} \omega_i a_1^{in}, \sum_{i=1}^{k} \omega_i a_2^{in}, \sum_{i=1}^{k} \omega_i a_L^{in}, \sum_{i=1}^{k} \omega_i a_R^{in}\right), n = \overline{1,N}\right\}. \quad (4.30)$$

Let us prove fulfillment of the condition (4.26), which ensures a Pareto optimality of generalized formalized result of expert evaluations of qualitative characteristic of an object group. Since

$$\min_{i=1,k}\left[\mu_1^n(x), \mu_2^n(x), ..., \mu_k^n(x)\right] =$$

$$\sum_{i=1}^{k} \omega_i \left\{\min_{i=1,k}\left[\mu_1^n(x), \mu_2^n(x), ..., \mu_k^n(x)\right]\right\} \leq \sum_{i=1}^{k} \omega_i \mu_i^n(x) =$$

$$= \mu^n(x) \leq \sum_{i=1}^{k} \omega_i \left\{\max_{i=1,k}\left[\mu_1^n(x), \mu_2^n(x), ..., \mu_k^n(x)\right]\right\} =$$

$$= \max_{i=1,k}\left[\mu_1^n(x), \mu_2^n(x), ..., \mu_k^n(x)\right], \ \forall x \in [0,1], \ n = \overline{1,N}, \quad (4.31)$$

Then we obtain

$$\min_{i=1,k}\left[\mu_1^n(x), \mu_2^n(x), ..., \mu_k^n(x)\right] \leq \mu^n(x) \leq$$

$$\leq \max_{i=1,k}\left[\mu_1^n(x), \mu_2^n(x), ..., \mu_k^n(x)\right], \ \forall x \in [0,1], \ n = \overline{1,N}.$$

Thus, the generalized result of expert evaluations of qualitative characteristic of an object group constructed within the limits of set Θ^k elements is Pareto optimal one.

In order to identify fuzzy evaluations of appearance of qualitative characteristic for real objects with one of linguistic values X_l, $l = \overline{1,m}$ of an estimated attribute, it is necessary to compare membership functions $\mu^n(x)$, $n = \overline{1,N}$ with membership functions $f_l(x)$, $l = \overline{1,m}$ of terms X_l, $l = \overline{1,m}$ of the generalized model constructed within the limits of set Ξ^k elements. To compare membership functions $\mu^n(x)$, $n = \overline{1,N}$ with membership functions $f_l(x)$, $l = \overline{1,m}$, the indexes can be used, without limiting generality,

$$\lambda_l^n = \frac{\int\limits_0^1 \min\left[\mu^n(x), f_l(x)\right]dx}{\int\limits_0^1 \max\left[\mu^n(x), f_l(x)\right]dx}, \quad l = \overline{1,m}, \quad n = \overline{1,N}.$$

or

$$\sigma_l^n = \int\limits_0^1 \left|\mu^n(x) - f_l(x)\right|dx, \quad l = \overline{1,m}, \quad n = \overline{1,N}.$$

Let us denote a possibility that fuzzy number with membership function $\mu^n(x)$ is equal to fuzzy number $\tilde{X}_j$ with membership function $f_j(x)$ with $\text{Pos}\left[\mu^n(x) = f_j(x)\right]$. As known [153],

$$\text{Pos}\left[\mu^n(x) = f_j(x)\right] = \max_x \min\left[\mu^n(x), f_j(x)\right].$$

If $\text{Pos}\left[\mu^n(x) = f_j(x)\right] = \gamma$, the evaluation of appearance of qualitative characteristic of n-th object (with membership function $\mu^n(x)$, $n = \overline{1,N}$) is identified using linguistic value X_j with possibility γ.

4.6 A Method of Determination of Optimum Sets of Characteristic Linguistic Values

Experts can apply different sets of their linguistic values to estimate or describe characteristics. Some sets bring difficulties to experts due to the insufficiency of values, and other sets — due to redundancy of values. As a result of these difficulties one should expect growth of fuzziness and mismatch of the information provided by experts.

While estimating of attributes' appearances by an expert, a natural problem is: "What are criteria to choose an optimum range of a linguistic scale applied to estimate any characteristic?" In [28], the following criteria of an optimality of set of characteristic linguistic values are defined:

1) set of values used by experts provides minimum uncertainty while describing real objects;
2) set of values which ensure maximum consistency of the expert information.

In [28], problem of definition of optimum sets of characteristic linguistic values is solved only if the first criterion is satisfied. In the present paragraph, while defining the optimum sets of characteristic linguistic values the maximin problem is solved.

For the purpose of the characteristic X evaluation or description, k experts are involved.

The sets

$$T_1 = \{X_1, X_2\}, T_2 = \{Y_1, Y_2, Y_3\}, ..., T_{n-1} = \{Z_1, Z_2, ..., Z_n\}$$

of characteristic X linguistic to be applied to estimate the characteristic considered, are formulated. After that, the experts are offered to estimate (or to describe) this characteristic sequentially within the limits of each formulated sets of its linguistic values.

Let

$$T_1 = \{X_1, X_2\}, T_2 = \{Y_1, Y_2, Y_3\}, ..., T_{n-1} = \{Z_1, Z_2, ..., Z_n\}$$

are COSS term-sets constructed within the limits of these sets based on the information obtained from each of k experts. Let us denote a model of expert evaluations of characteristic X by p-th expert within the limits of term-set T_i (COSS of p-th expert with a term-set T_i), with

$$P_i^p, \ i = \overline{1, n-1}, \ p = \overline{1, k}$$

and the generalized model of expert evaluations of characteristic X within the limits of term-set T_i with P_i, $i = \overline{1, n-1}$. Let us denote fuzziness degree of the model P_i with $\xi(T_i)$, $i = \overline{1, n-1}$, and index of the general consistency of models P_i^p, $p = \overline{1, k}$ with k_i.

For a consistency index, we construct COSS with universal set [0.1], terms "low", "high" and membership functions of terms $\mu_1(x), \mu_2(x)$, without limiting a generality, for example, $\mu_1(x) \equiv (0; 0,25; 0; 0,50)$, $\mu_2(x) \equiv (0,75; 1; 0,25; 0)$. For a fuzziness degree we construct COSS with universal set [0; 0.5], terms "small", "big" and membership functions of terms $\eta_1(x), \eta_2(x)$, without limiting a generality, for example, $\eta_1(x) \equiv (0; 0,20; 0; 0,20)$, $\eta_2(x) \equiv (0,40; 0,50; 0,20; 0)$.

Let us calculate, within the limits of all sets of linguistic values, the characteristic of membership value of fuzziness degrees of the generalized models to the term "small" — $\eta_1[\xi(T_i)]$, $i = \overline{1, n-1}$, and membership values of consistency indexes of experts' models to the term "high" — $\mu_2(k_i)$, $i = \overline{1, n-1}$.

Let us define

$$\theta_i = \min\{\eta_1[\xi(T_i)], \ \mu_2(k_i)\}, \ i = \overline{1, n-1}.$$

Then, the set of characteristic linguistic values is considered an optimum set (Fig. 4.2), if

$$\theta_j = \max_{1 \le i \le n-1} \theta_i.$$

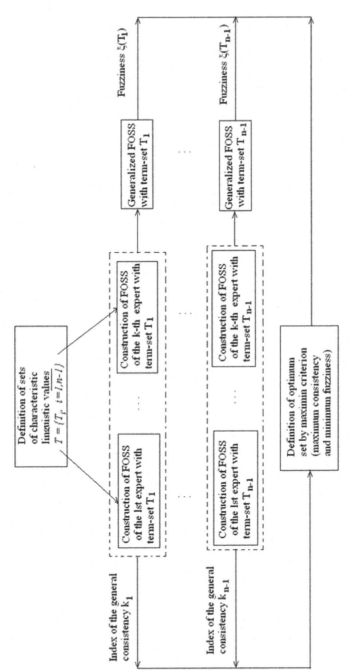

Fig. 4.2 The scheme of definition of optimum sets of characteristic linguistic values

4.7 Examples of Practical Application of the Developed Methods

Example 4.1. Building of the generalized result of expert evaluations of educational literature quality. Let us consider conditions of the example 3.3 and the related results obtained. Based on the analysis described in that example, the conclusion is drawn that the result of an evaluation of the educational literature quality made by the third expert essentially differs from results of evaluation of the educational literature quality made by other experts. This conclusion is validated with the fact that the result system excluding the results provided by the third expert has an additive index of the general consistency 0.266, and multiplicative index of the general consistency 0.252. Considering that results of experts are fractionalized onto two groups of results similar to each other, let us determine consistency indexes of two subsystems {2, 4, 5}, {1, 3, 6}. The first subsystem has additive and multiplicative indexes 0.522 and 0.5204, accordingly, and the second subsystem - 0.503 and 0, accordingly.

According to the research carried out, the system of results of the second, fourth and fifth experts is considered, and within the limits of their results the generalized result of the evaluation is determined. By definition of the experts, weight coefficients are equal to following values: $\omega_2 = 1/2$, $\omega_4 = 1/3$, $\omega_5 = 1/6$.

The generalized result of an evaluation of quality of the educational literature is defined as a linear combination of results of the second, fourth and fifth experts. This result is summarized in Table 4.1 in the form of parameters of membership functions of fuzzy evaluations of each object, the accurate evaluations obtained as a result of defuzzification of fuzzy evaluations by a method of a gravity, and levels of a verbal scale assigned to each object.

Example 4.2. Building of the generalized expert approach to an evaluation of a condition of "large-leaved linden" plants. Let us consider conditions of an example 3.4 and define weight coefficients of expert results within the limits of similarity relation $\hat{R}_2$. With the confidence level equal to 0.684, expert results enter the same cluster, therefore $\omega_i = 1/3$, $i = \overline{1,3}$. With the confidence level equal to 0.685, results of the second and third experts enter one cluster, therefore their weight coefficients are equal: $\omega_2 = \omega_3 = 5/12$. The weight coefficient of result of the first expert ω_1 is equal to 1/6. With the confidence level equal to unity, $\breve{\rho}_1 = 2,319$, $\breve{\rho}_2 = 2,369$, $\breve{\rho}_3 = 2,32$, therefore $\omega_1 = 1/6$, $\omega_2 = 1/2$, $\omega_3 = 1/3$ (Table 4.2).

As said above, weight coefficients of expert results within the limits of similarity relation $\hat{R}_1$ coincide with weight coefficients of similarity relation $\hat{R}_2$. With that, confidence levels are equal 0.857; 0.863; 1, accordingly.

Table 4.1 Formalized generalized result of the evaluation

μ	M	Defuzzification M	Scale level
μ_1	(0,6790;0,7630;0,1750;0,1580)	0,7158	5
μ_2	(0,4380;0,4545;0,1750;0,1750)	0,4462	4
μ_3	(0;0,0457;0;0,0916)	0,0469	1
μ_4	(0,6790;0,7630;0,1750;0,1580)	0,7158	5
μ_5	(0,1798;0,1880;0,01085;0,1250)	0,1893	2
μ_6	(0,5040;0,1580;0,1750)	0,5040	4
μ_7	(0,0623;0,1125;0,0415;0,1250)	0,1127	2
μ_8	(0,9210;1;0,1580;0)	0,9171	6
μ_9	(0,57590;0,6225;0,1750;1830)	0,6047	5
μ_{10}	(0,2875;0,3290;0,1250;0,1750)	0,3240	3
μ_{11}	(0,1373;0,1625;0,0916;0,1250)	0,1605	2
μ_{12}	(0,6790;0,7630;0,1750;0,1580)	0,7158	5
μ_{13}	(0,9210;1;0,1580;0)	0,9171	6
μ_{14}	(0,6295;0,7053;0,1750;0,1415)	0,6571	5
μ_{15}	(0,1803;0,2218;0,0755;0,1584)	0,2267	2
μ_{16}	(0,5040;0,1580;0,1750)	0,5040	4
μ_{17}	(012450;0,3035;0,1080;0,1750)	0,2949	3
μ_{18}	(0,8658;0,9745;0,1665;0,1700)	0,8769	6
μ_{19}	(0,8658;0,9745;0,1665;0,1700)	0,8769	6
μ_{20}	(0,2450;0,3035;0,1080;0,1750)	0,2949	**3**

Table 4.2 Weight coefficients of expert results within the limits of similarity relation $\hat{R}_2$

Confidence level	Weight coefficients
0.684	$\omega_1 = 1/3, i = \overline{1,3}$
0.685	$\omega_1 = 1/6, \omega_2 = \omega_3 = 5/12$
1	$\omega_1 = 1/6, \omega_2 = 1/2, \omega_3 = 1/3$

Let us define the formalized generalized approach to an evaluation of the "large-leaved linden" plants state: $\omega_1 = 1/6$, $\omega_2 = 1/2$, $\omega_3 = 1/3$. Based on the values of weight coefficients of expert results, let us define the formalized generalized approach of experts to an evaluation of the "large-leaved linden" plants state, which has the following membership functions:

$$f_1(x) \equiv (0; 0,019; 0; 0,0125);$$
$$f_2(x) \equiv (0,0315; 0,0125; 0,0125);$$
$$f_3(x) \equiv (0,044; 0,0565; 0,0125; 0,0375);$$
$$f_4(x) \equiv (0,094; 0,0982; 0,0375; 0,0458);$$
$$f_5(x) \equiv (0,144; 0,271; 0,0458; 0,3);$$
$$f_6(x) \equiv (0,571; 0,7938; 0,3; 0,1375);$$
$$f_7(x) \equiv (0,9313; 1; 0,1375; 0).$$

Example 4.3. Building of an optimum linguistic scale for the software completeness evaluation. For defining the optimum set of the linguistic scale used to estimate the software completeness, five experts generated three ranges of this scale:

1. "incompleteness", "partial completeness", "complete completeness";
2. "incompleteness", "partial completeness", " basic completeness", "complete completeness":
3. "incompleteness", "partial completeness", "basic completeness", "essential completeness", "complete completeness".

To estimate a characteristic "complete completeness" which characterizes degree of the software possession, all necessary parts and features required for carrying out its explicit and implicit functions, 25 software products for supporting financial activity of firms was chosen (Table 4.3).

Results of the software completeness evaluation made by five experts within the limits of a linguistic scale "incompleteness", "partial completeness", "basic completeness", "complete completeness" are summarized in Table 4.4.

Based on the method described in § 2.2 and data of Table 4.4, membership functions of term-sets of five COSS's (models of expert evaluations of software completeness) the graphs of which are represented in Fig. 4.3, were obtained.

Using definitions of §3.1, an additive index κ of the general consistency of those models and the multiplicative index $\widetilde{\kappa}$ of the general consistency of models was calculated, $\kappa = 0,705$, $\widetilde{\kappa} = 0,695$ respectively. Based on the obtained values the conclusion was drawn on a sufficient consistency of models of expert evaluations of software completeness.

Table 4.3 Software

Item No.	Software	Manufacturer
1	KDCalc Java	Knowledge Dynamics
2	KDCalcNET	Knowledge Dynamics
3	MStockTA	MSoftDevelopment
4	NumberToWords	Total Technology
5	ACCPAS	ActiveWebSoftwares

Table 4.3 (*continued*)

6	AFD BankFinder	AFD Software
7	Check Writer	Database Creations
8	A1BACSAX	Al Computer Software
9	A1 UK Bank Account Validation	Al Computer Software
10	Bonds for NET	WebCab
11	Cashflow	Vercom Systems
12	Approved List Manager	Procon Software and Support
13	Loan Calculator	AJE Components
14	Loan Engine	AJE Components
15	QBIIFUI	Massinissa Software
16	EasyTax Professional	Terra Base
17	FinLib	TeraTech
18	Positively Business Point of Sale	Database Creations
19	Plain OprionSolver Dev	Derivicom
20	Plain OptionSolver XL	Derivicom
21	Portfolio for NET	WebCab
22	Stores Manager	Procon Software and Support
23	Tenders Manager	Procon Software and Support
24	WebCab Equities. Interest and Real estate (J2SE Edition)	WebCab
25	WebCab Options and Futures for NET	WebCab

Table 4.4 Expert evaluations of software completeness

Number of the expert	Incompleteness	Partial completeness	Basic completeness	Complete completeness
		Initial results		
1	4	10	8	3
2	3	12	8	2
3	3	10	8	4
4	4	9	9	3
5	3	7	10	5
		Relative results		
1	0,16	0,4	0,32	0,12
2	0,12	0,48	0,32	0,08
3	0,12	0,4	0,32	0,16
4	0,16	0,36	0,36	0,12
5	0,12	0,28	0,4	0,2

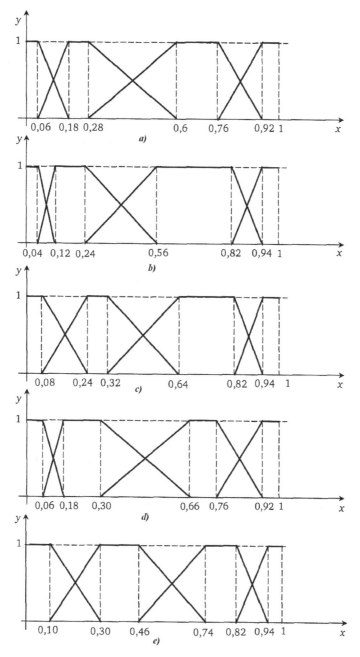

Fig. 4.3 Membership functions of terms of expert models; a — the first expert; b — the second expert; c — the third expert; d — the fourth expert; e — the fifth expert

Table 4.5 Elements of a matrix and transitive closure of the expert models pairwise consistency matrix

Matrix elements

1	0,819	0,875	0,923	0,792
0,819	1	0,812	0,834	0,768
0,875	0,812	1	0,938	0,884
0,923	0,834	0,938	1	0,858
0,792	0,768	0,884	0,858	1

Elements of transitive closures

1	0,834	0,923	0,923	0,884
0,834	1	0,834	0,834	0,834
0,923	0,834	1	0,938	0,884
0,923	0,834	0,938	1	0,884
0,884	0,834	0,884	0,884	1

Indexes of pairwise consistency of expert models (Table 4.5) were calculated according to the definition (see §3.1).

As one can see from this matrix, models of the third and fourth experts have the greatest index of pairwise consistency.

Further, the fuzzy binary relation of similarity based on the computed indexes of pairwise consistency of expert models is constructed. The matrix of pairwise consistency of expert models is not transitive; therefore its transitive closure (see the bottom part of Table 4.5) is determined.

Table 4.6 Fuzzy clusterization of expert models under similarity relation $\hat{R}$

Confidence level	Cluster
0,834	{1,2,3,4,5}
0,884	{1,3,4,5},{2}
0,923	{1,3,4},{2},{5}
0,938	{3,4},{1},{2},{5}
1	{1},{2},{3},{4},{5}

This matrix defines using set of expert models the fuzzy relation of similarity $\hat{R}$ with fuzziness degree equal to 0.196. Relation $\hat{R}$ is decomposed onto equivalence relations, the results obtained are given in Table 4.6, and weight coefficients (see §4.2) of expert models (Table 4.7) are determined based on the similarity $\hat{R}$.

Fuzziness degrees of expert models for software completeness evaluations are determined: $\zeta_1 = 0,30$, $\zeta_2 = 0,26$, $\zeta_3 = 0,30$, $\zeta_4 = 0,32$, $\zeta_5 = 0,30$.

Table 4.7 Weight coefficients of models on the basis of similarity relation $\hat{R}$

Confidence level	Weight coefficients
0834	$\omega_1 = 1/5m, i = \overline{1,5}$
0.884	$\omega_1 = \omega_3 = \omega_4 = \omega_5 = 7/30, \omega_2 = 1/5$
0.923	$\omega_1 = \omega_3 = \omega_4 = 4/15, \omega_5 = 2/15, \omega_2 = 1/15$
0.938	$\omega_3 = \omega_4 = 3/10, \omega_1 = 1/5, \omega_5 = 2/15, \omega_2 = 1/15$
1	$\omega_3 = 4/15, \omega_4 = 1/3, \omega_1 = 1/5, \omega_5 = 2/15, \omega_2 = 1/15$

Within the limits of these results weight coefficients (see §4.3) are obtained: $\omega_2 = 1/3$, $\omega_1 = \omega_3 = \omega_5 = 1/5$, $\omega_4 = 1/15$.

Using results described in §4.1, it is possible to construct the generalized model of expert evaluations of software completeness in the following form:

$$X = \left\{ f_l(x), l = \overline{1,4} \right\} = \left\{ \sum_{i=1}^{5} \omega_i \mu_{il}(x), l = \overline{1,4} \right\};$$

$$f_l(x) \equiv \left(a_1^l, a_2^l, a_L^l, a_R^l \right) = \left\{ \sum_{i=1}^{5} \omega_i a_1^{il}(x), \sum_{i=1}^{5} \omega_i a_2^{il}, \sum_{i=1}^{5} \omega_i a_L^{il}, \sum_{i=1}^{5} \omega_i a_R^{il}, l = \overline{1,4} \right\}.$$

Taking into account the weight coefficients defined within the limits of similarity relation, it is possible to obtain following membership functions of generalized model X^1:

$$f_1^1(x) \equiv (0; 0{,}066; 0; 0{,}132); \quad f_2^1(x) \equiv (0{,}198; 0{,}318; 0{,}132; 0{,}328);$$
$$f_3^1(x) \equiv (0{,}646; 0{,}788; 0{,}328; 0{,}141); \quad f_4^1(x) \equiv (0{,}929; 1; 0{,}141; 0).$$

While using the weight coefficients constructed within the limits of computed fuzziness degrees of expert models, the following membership functions of generalized model X^2 are obtained:

$$f_1^2(x) \equiv (0; 0{,}065; 0; 0{,}130); \quad f_2^2(x) \equiv (0{,}195; 0{,}312; 0{,}130; 0{,}314);$$
$$f_3^2(x) \equiv (0{,}626; 0{,}804; 0{,}314; 0{,}130); \quad f_4^2(x) \equiv (0{,}934; 1; 0{,}130; 0).$$

Constructed generalized models X^1 and X^2 are Pareto optimal and optimal as per the method of the least weighed quadrates of differences between parameters of these models and parameters of expert models.

Using results of §4.4, it is possible to construct generalized model X^3 of expert evaluations of software completeness, which keeps a maxima of the information obtained from all experts:

$$f_1^3(x) \equiv (0; 0,067; 0; 0,137); \quad f_2^3(x) \equiv (0,204; 0,312; 0,137; 0,337);$$
$$f_3^3(x) \equiv (0,649; 0,745; 0,337; 0,170); \quad f_4^3(x) \equiv (0,915; 1; 0,170; 0).$$

In § 4.4 it was underlined that from the point of view of information losses, optimal generalized model does not generally satisfy to Pareto condition. In this specific case while comparing X^3 with models of all experts it becomes clear that generalized model X^3 satisfies to the Pareto condition. Information loss is equal to $\sigma_{\text{опт}} = 0,095$ while constructing the model X^3.

Comparison of information losses σ_1 and σ_2 while constructing the generalized models X^1 and X^2, accordingly, has shown that $\sigma_1 = 0,106$, $\sigma_2 = 0,118$. Fuzziness of generalized models X^1, X^2, X^3 are equal to 0.301; 0.287; and 0.322, accordingly. Analogues of reliability index (see §2.7) of these models are equal 0.114; 0.107 and 0.1685, accordingly.

With the results of the research carried out, the generalized model of expert evaluations of software completeness was represented with model X^1 selected, because it is Pareto optimal one, information loss is close to optimum when it was constructed, and values of fuzziness degree of X^1 and analogue of a reliability index occupies intermediate position between values of fuzziness degrees and analogues of reliability index for models X^2 and X^3.

The researches carried out within the limits of the scale "incompleteness", "partial completeness", "complete completeness" allowed obtaining the generalized model X^4 with membership functions

$$f_1^4(x) \equiv (0; 0,060; 0; 0,261); \quad f_2^4(x) \equiv (0,321; 0,378; 0,261; 0,432);$$
$$f_3^4(x) \equiv (0,810; 1; 0,432; 0).$$

The additive index to the general consistency of expert models is equal to 0.594, the multiplicative index $\tilde{\kappa}$ is equal to 0.563, fuzziness degree of the model X^4 is equal to 0.347.

While using the scale «complete completeness», «partial completeness», «basic completeness», «essential completeness», «complete completeness», the generalized model X^5 was obtained with membership functions

$$f_1^5(x) \equiv (0; 0,061; 0; 0,117); \quad f_2^5(x) \equiv (0,178; 0,246; 0,117; 0,212);$$
$$f_3^5(x) \equiv (0,458; 0,645; 0,212; 0,172);$$
$$f_4^5(x) \equiv (0,817; 0,825; 0,172; 0,162); \quad f_5^5(x) \equiv (0,987; 1; 0,162; 0).$$

The additive index κ to the general consistency of expert models is equal to 0.604, and the multiplicative index $\tilde{\kappa}$ is equal to 0.582, fuzziness degree of the model X^5 is equal to 0.331.

On the basis of comparison of the results obtained within the limits of three formulated sets of linguistic values of software quality characteristic «complete completeness», the conclusion is drawn that optimum set of a linguistic scale for an expert evaluation of software completeness is the set of terms: «complete completeness», «partial completeness», «basic completeness», «complete completeness».

Chapter 5
Ratings of Objects

5.1 Obtaining of Rating Points of Objects within the Limits of a Qualitative Characteristic

Rating points systems are widely used in various human activities (educational process, economics, techniques, etc.) and are of great importance in decision making problems [154-166]. These systems make it possible to get available and timely information in the form of an aggregative index and to use it in decision making problems. The complexity of obtaining rating points for objects with qualitative characteristics results from the general complexity of the quantitative assessment of qualitative characteristics. This complexity is also associated with the necessity of taking into account characteristics and judgements of the surveyors who take decisions based on their personal assessment. As a rule, the qualitative characteristics are scored in different scales and are often incomparable in principle. The elements of these scales (as a rule, order-type scales) are transformed into scores. Such transformation needs some substantiation because stability of the final findings depends on it.

The following example will make the point clear. Let us suppose that two objects got 4 and 3 points for one characteristic and 4 and 5 points for the other characteristic correspondingly. As a result of two assessments each object gets the same total score that equals 8. The conclusion is made that they have similar rating points and similar rating correspondingly. Since we deal with the order-type scale while assessing objects' qualitative characteristics, we shall apply strictly increasing transformation Φ of this scale, that is acceptable: $\Phi(3) = 3, \Phi(4) = 4, \Phi(5) = 7$. It is known [1] that an acceptable transformation of the values of the assessed quality feature is such a transformation that retains subject matter of the type of assessment involved. In accordance with the transformation applied the total score remained the same for object 1 while it changed for object 2 and has become equal 10 points. Thus the rating point of the second object has increased. The stability of the results after the acceptable transformation is violated that testifies to the fact that transformation of verbal scales' elements into scores needs some substantiation.

Problems of obtaining ratings and rating points of real objects arise while estimating the objects considered within the limits of both quantitative and qualitative characteristics.

O.M. Poleshchuk and E.G. Komarov: Expert Fuzzy Info. Processing, STUDFUZZ 268, pp. 147–182.
springerlink.com

Obtaining of ratings within the limits of quantitative characteristics [167] is often enough reduced to determination of scalar integral indexes using vectors of partial quantitative evaluations.

Methods of obtaining ratings within the limits of qualitative characteristics [21—27. 168—170] use the approach [167] applied to qualitative characteristic evaluation. However, this approach has some restrictions connected with peculiarities of scales used to measure qualitative characteristics, those restrictions are considered in detail in § 1.2.

While estimating of qualitative characteristics verbal scales are used rather often, with numerical points, or elements of ordinal scales, put in correspondence to levels of these scales. In the expert evaluation theory, the problem of determination of the numerical point collections put in correspondence to elements of ordinal scales, is one of main problems [171]. In case of use of an unjustified arbitrary collection of numerical points the stability of definitive conclusions is broken. Thus, either the choice of a collection of the numerical points put in correspondence to levels of verbal scales should be justified within the limits of each specific task, or a new approach (excluding necessity of operation with a particular collection of numerical points) for determination of rating results should be offered.

Methods [172—174] to obtain rating points of objects and groups of objects by one and several qualitative and quantitative characteristics based on their linguistic values are described below. These methods allow operating with abstract values, namely, values of membership functions of linguistic values of estimated characteristics rather than with numerical points.

Let us consider procedure of qualitative characteristic evaluation be simple, if an expert refers an object to a certain verbal level of characteristic manifestation and gives certain points corresponding to this level. For example, an expert estimates convenience of software interface within the limits of a verbally-numerical scale with four levels of a state: "inconvenient" — 1 point, "inconvenient enough" — 2 points, "convenient enough" — 3 points, "convenient" — 4 points. Giving certain points, the expert thereby defines a rating point for each object. However it is necessary to note, that rather often such simple procedures of qualitative characteristic evaluation give rough and approximate estimate which can be erroneous, therefore these procedures need to be sophisticated.

Let us consider procedure of qualitative characteristic evaluation be complex, if it consists of an evaluation of several sub-characteristics composing this characteristic. An example of complex procedure of qualitative characteristic evaluation is a procedure used to evaluate knowledge of students trained within the limits of a certain subject by following components: academic achievements during a semester, testing results on different sections of the subject, participation in research work within the limits of this subject etc. The result of a complex evaluation of qualitative characteristic can be presented in the form of a vector with co-ordinates being intermediate evaluations of the sub-characteristics. Then, a problem of aggregation of separate evaluations into a uniform integral (rating) evaluation and assignment of the accepted or specially developed gradation (verbal levels) to objects within the limits of the characteristic considered.

In this paragraph we offer to construct convolution of comparable components of a membership of these evaluations to the fuzzy numbers which formalize linguistic values of estimated qualitative characteristic instead of convolution of the intermediate evaluations obtained.

Let us consider a group N of objects which are estimated within the limits of manifestation of a qualitative characteristic X. Rating points are defined on the basis of mark evaluations k for sub-characteristics, which compose the characteristic. A minimum quantity of points with which an object can be estimated within the scope of i-th sub-characteristic is equal to zero, and a maximum quantity of points is equal to Z_i, $i = \overline{1,k}$.

Let us denote an evaluation of n-th object $n = \overline{1,N}$ in the scope of i-th sub-characteristic $i = \overline{1,k}$ with z_i^n. Let us normalize evaluations z_i^n, $i = \overline{1,k}$ and present result of the evaluation of n-th object in the form of the vector

$$\left(\frac{z_1^n}{Z_1}, \frac{z_2^n}{Z_2}, ..., \frac{z_k^n}{Z_k}\right) = \left(m_1^n, m_2^n, ..., m_k^n\right), \ n = \overline{1,N}. \tag{5.1}$$

Let X_l, $l = \overline{1,m}$ be levels of a verbal scale arranged by increase of intensity of characteristic X and applied to the estimate purpose. Within the limits of the building method (see § 2.2) of COSS with title X and membership functions $\mu_l(x)$, $l = \overline{1,m}$ of terms X_l, $l = \overline{1,m}$ is carried out. The fuzzy numbers corresponding to terms X_l are denoted with $\tilde{X}_l$. The amount of terms is defined by an amount of accepted (or specially developed) verbal levels of intensity of characteristic manifestations X. According to the requirements to membership functions of COSS terms, for vector co-ordinates (5.1) validity of the following equalities follows:

$$m_i^n = m_i^n \cdot \left(\sum_{l=1}^{m} \mu_l\left(m_i^n\right)\right) = \sum_{l=1}^{m} m_i^n \mu_l\left(m_i^n\right), \ i = \overline{1,k}. \tag{5.2}$$

Considering a fuzziness attributed to evaluation procedure, let us in (5.2) substitute normalized evaluations m_i^n in each product $m_i^n \mu_l\left(m_i^n\right)$, $l = \overline{1,m}$ with fuzzy numbers $\tilde{X}_l$, $l = \overline{1,m}$, accordingly. Such procedure is referred to as fuzzifying of definite data [53]. Then, vector co-ordinates of evaluations of n-th object, $n = \overline{1,N}$ are fuzzy numbers

$$\tilde{m}_i^n = \mu_1\left(m_i^n\right) \otimes \tilde{X}_1 \oplus ... \oplus \mu_m\left(m_i^n\right) \otimes \tilde{X}_m, \ i = \overline{1,k}. \tag{5.3}$$

Parameters of these numbers are defined by component-wise multiplication of parameters of fuzzy numbers $\tilde{X}_l$, $l \equiv \left(a_1^l, a_2^l, a_L^l, a_R^l\right)$, $l = \overline{1,m}$ by usual numbers $\mu_l\left(m_i^n\right)$, $l = \overline{1,m}$, $i = \overline{1,k}$, $n = \overline{1,N}$ and their subsequent addition, i.e. for (5.3) we obtain

$$\tilde{m}_i^n \equiv \left(\sum_{l=1}^m \mu_l \left(m_i^n \right) a_1^l, \sum_{l=1}^m \mu_l \left(m_i^n \right) a_2^l, \sum_{l=1}^m \mu_l \left(m_i^n \right) a_L^l, \sum_{l=1}^m \mu_l \left(m_i^n \right) a_R^l \right). \qquad (5.4)$$

The vector of evaluations of n-th object (5.1) is replaced with a new vector of fuzzy evaluations:

$$\left(\tilde{m}_1^n, \tilde{m}_2^n, ..., \tilde{m}_k^n \right), \ n = \overline{1, N}. \qquad (5.5)$$

Each evaluation shall contribute to a total rating point with some weight coefficient which is determined based on importance of a stage related to the evaluation. With the system of preferences not available, stages and related evaluations are considered equivalent with weights $\omega_i = 1/k$, $i = \overline{1,k}$. With the system of preferences is available, we range stages, each of them aimed at measuring of one sub-characteristic, in the order of decrease of their contribution to the total evaluation. Let us use Fishburn scale and determine weight coefficients of sub-characteristics as follows:

$$\omega_i = \frac{2(k-i+1)}{k(k+1)}, \ i = \overline{1,k}.$$

Other approaches to determination of weight coefficients are described in [175].

Let us determine fuzzy rating of characteristic manifestation X for n-th object by the formula

$$\tilde{A}_n = \omega_1 \otimes \tilde{m}_1^n \oplus ... \oplus \omega_k \otimes \tilde{m}_k^n, \ n = \overline{1, N} \qquad (5.6)$$

or

$$\tilde{A}_n \equiv \left[\sum_{i=1}^k \omega_i \sum_{l=1}^m \mu_l \left(m_i^n \right) a_1^l, \sum_{i=1}^k \omega_i \sum_{l=1}^m \mu_l \left(m_i^n \right) a_2^l; \right.$$
$$\left. \sum_{i=1}^k \omega_i \sum_{l=1}^m \mu_l \left(m_i^n \right) a_L^l, \sum_{i=1}^k \omega_i \sum_{l=1}^m \mu_l \left(m_i^n \right) a_R^l \right] \equiv \left(\Delta_1^n, \Delta_2^n, \Delta_L^n, \Delta_R^n \right).$$

Let us define a confidential interval for definite rating x_n, which characterizes characteristic manifestation X of n-th object, $n = \overline{1, N}$. From definition of level sets for fuzzy number it follows that for confidence level $\eta_n(x_n) \geq \alpha$, $0 < \alpha < 1$ rating x_n of characteristic manifestation X for n-th object, $n = \overline{1, N}$ is within an interval

$$\Delta_1^n - \Delta_L^n L^{-1}(\alpha) \leq x_n \leq \Delta_2^n + \Delta_R^n R^{-1}(\alpha), \qquad (5.7)$$

where η_n is membership function of fuzzy number $\tilde{A}_n$, $n = \overline{1, N}$.

Let us defuzzificate fuzzy numbers $\tilde{A}_n$, $\tilde{X}_1$, and $\tilde{X}_m$ using gravity method (1.8). It is resulted in definite numbers which will be denoted with A_n, $n = \overline{1,N}$, B_1, B_m, accordingly.

Number A_n, $n = \overline{1,N}$ is referred to as point-wise rating of manifestation of qualitative characteristic X for n-th object, $n = \overline{1,N}$.

Let us compute normalized rating by the formula

$$E_n = \frac{A_n - B_1}{B_m - B_1}. \tag{5.8}$$

Evaluation E_n is referred to as a degree of characteristic X manifestation intensity for n-th object. E_n range is a segment [0.1].

Thus, the method allows quantitative evaluations of qualitative characteristics' manifestations.

If it is necessary to assign one of qualification levels (one of linguistic values X_l, $l = \overline{1,m}$ of characteristic X) to n-th object, we can use the following indexes, without limiting a generality,

$$\lambda_n^l = \frac{\int\limits_0^1 \min[\eta_n(x), \mu_l(x)]dx}{\int\limits_0^1 \max[\eta_n(x), \mu_l(x)]dx}, \quad l = \overline{1,m}, \ n = \overline{1,N} \tag{5.9}$$

or

$$\sigma_n^l = \int\limits_0^1 |\eta_n(x) - \mu_l(x)|dx, \ l = \overline{1,m}, \ n = \overline{1,N}. \tag{5.10}$$

If $\lambda_n^j = \max \lambda_n^l$ $\left(\sigma_n^j = \max \sigma_n^l\right)$, the possibility of equality $\tilde{A}_n$ and $\tilde{X}_j$ — $\mathrm{Pos}\!\left(\tilde{A}_n = \tilde{X}_j\right)$ is calculated. According to [153]

$$\mathrm{Pos}\!\left(\tilde{A}_n = \tilde{X}_j\right) = \max_x \min[\eta_n(x), \mu_j(x)]$$

If $\mathrm{Pos}\!\left(\tilde{A}_n = \tilde{X}_j\right) = \gamma$, then qualification level X_j is assigned to n-th object with the possibility γ.

5.2 Determination of Object Ratings and Qualification Levels by Several Qualitative Characteristics Measured in Marks and Verbal Scales

Let us consider a group of objects which are estimated by several qualitative characteristics. To measure these characteristics ordinal scales are used. With mark evaluations assigned to objects of the group, problem to aggregate these evaluations in a certain uniform index arises rather often. This index is a final rating index of object within the limits of considered characteristics and is used, for example, to assign the object with qualification level available among the accepted ones [23, 25, 174—179]. Building of final ratings based on arithmetical operations is incorrect because of incomparability, in sense and content, of object evaluations within the limits of different qualitative characteristics.

Therefore, we offer to construct convolution of comparable abstract values of indexes membership to COSS terms as the final rating. Transformation of evaluations of characteristic manifestation to values of COSS membership functions corresponds to measurement of these characteristics within a uniform scale and ensures adequacy of model below for determination of the general ratings of units within the limits of several characteristics.

The characteristics measured in mark scales. Let qualitative characteristics Y_j, $j = \overline{1,k}$ be estimated for the group of N objects. To estimate these characteristics mark scales with elements $y_j = \overline{0,K_j}$, $j = \overline{1,k}$, accordingly, are used. Let us introduce normalized marks $s_j = y_j / K_j$, $y_j = \overline{0,K_j}$, $j = \overline{1,k}$. Thus, $0 \le s_j \le 1$, $j = \overline{1,k}$. Following the evaluation results of all characteristics, it is necessary to assign one of the accepted qualification levels X_l, $l = \overline{1,m}$ to the objects. Levels are arranged in ascending order of characteristic manifestation intensity degree. Let us construct COSS with terms X_l, $l = \overline{1,m}$ using the method described in §2.2. Relative contents of objects (probably, of certain ideal group) a priori set within the limits of each qualification level are taken as parameters necessary for model-building of COSS. For example, if qualification levels are assigned to the enterprise employees, necessary parameters for building of COSS can be defined on the basis of the selected vacancy jobs within the limits of each qualification level.

Let us denote membership functions of fuzzy numbers $\tilde{X}_l$ corresponding to terms X_l with $\mu_l(x) \equiv \left(a_1^l, a_2^l, a_L^l, a_R^l \right)$. Let for n-th object, $n = \overline{1,N}$, manifestations of characteristics Y, $j = \overline{1,k}$ are estimated by numbers y_j^n, $j = \overline{1,k}$, $n = \overline{1,N}$, or normalized numbers s_j^n, $j = \overline{1,k}$, $n = \overline{1,N}$, accordingly. Then $\mu_l\left(s_j^n \right)$, $j = \overline{1,k}$, $n = \overline{1,N}$ are membership degrees of normalized marks to fuzzy numbers $\tilde{X}_l$, $l = \overline{1,m}$.

Let us introduce the M_n matrix, $n = \overline{1,N}$, with columns being degrees of membership of normalized marks s_j^n, $j = \overline{1,k}$ of n-th object $n = \overline{1,N}$ to fuzzy numbers $\tilde{X}_l$, $l = \overline{1,m}$:

$$\mathrm{M}_n = \begin{pmatrix} \mu_1\left(s_1^n\right) & \mu_1\left(s_2^n\right) & \cdots & \mu_1\left(s_k^n\right) \\ \mu_2\left(s_1^n\right) & \mu_2\left(s_2^n\right) & \cdots & \mu_2\left(s_k^n\right) \\ \cdot & \cdot & \cdots & \cdot \\ \mu_m\left(s_1^n\right) & \mu_m\left(s_2^n\right) & \cdots & \mu_m\left(s_k^n\right) \end{pmatrix}. \tag{5.11}$$

Let $P = \left(\tilde{X}_1, \tilde{X}_2, ..., \tilde{X}_m\right)$ be a vector the co-ordinates of which be fuzzy numbers. Result of the generalized multiplication of this vector by a matrix (5.11) is the vector $\mathrm{H}_n = P \otimes \mathrm{M}_n$, $\mathrm{H}_n = \left(\tilde{A}_1^n, \tilde{A}_2^n, ..., \tilde{A}_k^n\right)$, $n = \overline{1, N}$ with co-ordinates in the form of fuzzy numbers

$$\tilde{A}_j^n = \mu_1\left(s_j^n\right) \otimes \tilde{X}_1 \oplus ... \oplus \mu_m\left(s_j^n\right) \otimes \tilde{X}_m, \quad j = \overline{1, k}, \ n = \overline{1, N}. \tag{5.12}$$

These fuzzy numbers are fuzzy evaluations of manifestation of qualitative characteristics Y, $j = \overline{1, k}$ for n-th object. Let us assign corresponding weight to each characteristic

$$\omega_j, \ j = \overline{1, k}, \ \sum_{j=1}^{k} \omega_j = 1.$$

Let us define the fuzzy rating corresponding to manifestation of estimated characteristic Y, $j = \overline{1, k}$ for n-th object as follows:

$$\tilde{A}_n = \omega_1 \otimes \tilde{A}_1^n \oplus ... \oplus \omega_k \otimes \tilde{A}_k^n. \tag{5.13}$$

From (5.12), (5.13) it follows, that

$$\tilde{A}_n \equiv \left[\sum_{j=1}^{k} \omega_j \sum_{l=1}^{m} \mu_l\left(s_j^n\right) a_1^l, \sum_{j=1}^{k} \omega_j \sum_{l=1}^{m} \mu_l\left(s_j^n\right) a_2^l, \right.$$
$$\left. \sum_{j=1}^{k} \omega_j \sum_{l=1}^{m} \mu_l\left(s_j^n\right) a_L^l, \sum_{j=1}^{k} \omega_j \sum_{l=1}^{m} \mu_l\left(s_j^n\right) a_R^l \right]. \tag{5.14}$$

Let us denote parameters of fuzzy number $\tilde{A}_n$, $n = \overline{1, N}$ in (5.14) with $\delta_1^n, \delta_2^n, \delta_L^n, \delta_R^n$ and membership function with $\eta_n(x)$, accordingly. Let us determine a confidential interval for the definite rating x_n which characterize manifestations of characteristics Y, $j = \overline{1, k}$ for n-th object. With confidence level $\eta_n(x_n) \geq \alpha$, $0 < \alpha < 1$, rating x_n of manifestations of characteristics Y, $j = \overline{1, k}$ for n-th object lies within the interval

$$\delta_1^n - \delta_L^n L^{-1}(\alpha) \leq x_n \leq \delta_2^n + \delta_R^n R^{-1}(\alpha). \tag{5.15}$$

Let us defuzzificate fuzzy numbers $\tilde{A}_n$, $n = \overline{1, N}$, $\tilde{X}_1$ and $\tilde{X}_m$ using gravity method (1.8). As a result, we obtain definite numbers to be denoted with $\tilde{A}_n$, $n = \overline{1, N}$, B_1, B_m, accordingly.

The number A_n, $n = \overline{1, N}$ is referred to as a pointwise rating of manifestation of characteristic Y, $j = \overline{1, k}$ for n-th object.

Let us compute normalized rating by the formula

$$E_n = \frac{A_n - B_1}{B_m - B_1}, \quad n = \overline{1, N}. \tag{5.16}$$

Let the evaluation E_n, $n = \overline{1, N}$ be referred to as average intensity degree of manifestation of characteristics Y_j, $j = \overline{1, k}$ for n-th object, $n = \overline{1, N}$. Range of E_n, $n = \overline{1, N}$ is a segment $[0, 1]$. Thus, the method allows to obtain quantitative evaluations of manifestations of several qualitative characteristics.

To assign one of qualification levels X_l, $l = \overline{1, m}$ to n-th object, it is necessary to identify fuzzy number $\tilde{A}_n$, $n = \overline{1, N}$ having membership function v_n, $n = \overline{1, N}$ with one of fuzzy numbers $\tilde{X}_l$, $l = \overline{1, m}$ with membership functions $\mu_l(x)$, $l = \overline{1, m}$. For this purpose, let us calculate identification indexes:

$$\lambda_n^l = \frac{\int_0^1 \min[\mu_l(x), v_n(x)]dx}{\int_0^1 \max[\mu_l(x), v_n(x)]dx}, \quad l = \overline{1, m}, \ n = \overline{1, N} \tag{5.17}$$

or

$$\sigma_n^l = \int_0^1 |v_n(x) - \mu_l(x)|dx, \ l = \overline{1, m}, \ n = \overline{1, N}. \tag{5.18}$$

If

$$\lambda_n^j = \max_l \lambda_n^l, \ \left(\sigma_n^j = \max_l \sigma_n^l\right),$$

then $\mathrm{Pos}(\tilde{A}_n = \tilde{X}_j)$ is calculated.

If $\mathrm{Pos}(\tilde{A}_n = \tilde{X}_j) = \gamma$, then qualification level X_j is assigned to n-th object with the possibility γ.

The described method of obtaining object ratings within the scope of several qualitative characteristics is applicable under condition of measurement of all indications in ordinal numerical scales. The example illustrating the described method is followed with the method of obtaining object ratings within the scope of the several qualitative characteristics measured in verbal scales.

Characteristics measured within verbal scales. Let us consider group of N objects, the intensity of manifestation of qualitative characteristics X_j, $j = \overline{1,k}$ of which is estimated. Let X_{lj}, $l = \overline{1, m_j}$ be levels of the verbal scales applied to estimate characteristics X_j, $j = \overline{1,k}$, accordingly. Levels are arranged in ascending order of intensity of manifestation of these characteristics.

Let us denote with a_i^j relative numbers of objects of the considered group, which are referred to level X_{lj} while estimating characteristic X_j.

Based on these data, let us construct k COSS's with names X_j, and term-sets X_{lj}. Let us denote membership function of fuzzy number $\tilde{X}_{lj}$ corresponding to l-th object of j-th COSS with $\mu_{lj}(x)$. Let us refer fuzzy numbers $\tilde{X}_{lj}$ or their membership function $\mu_{lj}(x)$ as object evaluations. Let us denote an evaluation of n-th object within the limits of characteristic X_j with $\tilde{X}_j^n$ and $\mu_j^n(x) \equiv \left(a_{j1}^n, a_{j2}^n, a_{jL}^n, a_{jR}^n \right)$, $n = \overline{1,N}$, $j = \overline{1,k}$. Fuzzy number $\tilde{X}_j^n$ with membership function $\mu_j^n(x)$ is equal to one of fuzzy numbers $\tilde{X}_{lj}$. Let us denote weight coefficients of estimated characteristics with

$$\omega_j, \ \sum_{j=1}^{n} \omega_j = 1.$$

The fuzzy rating of n-th object within the limits of characteristic X_j is defined in the form of fuzzy number

$$\tilde{A}_n = \omega_1 \otimes \tilde{X}_1^n \oplus \ldots \oplus \omega_k \otimes \tilde{X}_k^n \tag{5.19}$$

with membership function

$$\mu_n(x) \equiv \left(\sum_{j=1}^{k} \omega_j a_{j1}^n, \sum_{j=1}^{k} \omega_j a_{j2}^n, \sum_{j=1}^{k} \omega_j a_{jL}^n, \sum_{j=1}^{k} \omega_j a_{jR}^n \right). \tag{5.20}$$

Let us define a confidential interval for definite rating y_n characterizing manifestations of characteristics X_j for n-th object. With confidence level $\mu_n(y_n) \geq \alpha$, $0 < \alpha < 1$ rating y_n of manifestation of characteristics X_j for n-th object lies within the interval

$$\sum_{j=1}^{k} \omega_j a_{j1}^n - L^{-1}(\alpha) \sum_{j=1}^{k} \omega_j a_{jL}^n \leq x_n \leq \sum_{j=1}^{k} \omega_j a_{j2}^n + R^{-1}(\alpha) \sum_{j=1}^{k} \omega_j a_{jR}^n. \tag{5.21}$$

Let us defuzzificate fuzzy numbers $\tilde{A}_n$, $n = \overline{1, N}$, $\overline{B}_1 = \omega_1 \otimes \overline{X}_{11} \oplus \ldots \oplus \omega_k \otimes \overline{X}_{1k}$, $\tilde{B}_m = \omega_1 \otimes \tilde{X}_{m_1 1} \oplus \ldots \oplus \omega_k \otimes \tilde{X}_{m_k k}$ using gravity method (1.8), and let us denote the obtained definite numbers as A_n, $n = \overline{1, N}$, B_1, B_m.

The number $\tilde{A}_n$, $n = \overline{1, N}$ is referred to as a pointwise rating of manifestation of qualitative characteristics X_j for n-th object.

Let us determine normalized rating of n-th object by the formula

$$E_n = \frac{A_n - B_1}{B_m - B_1}. \qquad (5.22)$$

Let refer the evaluation E_n as an average intensity degree of manifestation of characteristics X_j for n-th object. Range of E_n is a segment [0, 1]. Thus, the method allows to determine quantitative evaluations of manifestations of several qualitative characteristics.

Let us assume that, by results of an evaluation of all characteristics, it is necessary to assign one of accepted qualification levels D_l, $l = \overline{1, m}$ to the objects. Levels are arranged in ascending order of their rating. Let us construct COSS with terms D_l, $l = \overline{1, m}$ using the method described in §2.2. relative contents of objects (probably, of a certain ideal group) a priori set within the limits of each qualification level are taken as parameters necessary for building of COSS's. Let us denote membership functions of fuzzy numbers $\tilde{D}_l$, corresponding to terms D_l, with $\eta_l(x)$. To assign one of qualification levels D_l to n-th object, it is necessary to identify fuzzy number with membership function $\mu_n(x)$ and with one of term-sets having membership functions $\eta_l(x)$. For this purpose let us calculate identification indexes:

$$\lambda_n^l = \frac{\int_0^1 \min[\eta_l(x), \mu_n(x)]dx}{\int_0^1 \max[\eta_l(x), \mu_n(x)]dx} \qquad (5.23)$$

or

$$\sigma_n^l = \int_0^1 |\eta_l(x) - \mu_n(x)|dx. \qquad (5.24)$$

If

$$\lambda_n^j = \max_l \lambda_n^l, \quad \left(\sigma_n^j = \max_l \sigma_n^l\right),$$

then, $\mathrm{Pos}\left(\tilde{A}_n = \tilde{D}_j\right)$ is calculated.

If $\text{Pos}\left(\tilde{A}_n = \tilde{D}_j\right) = \gamma$, qualification level D_j is assigned to n-th object with a possibility γ.

5.3 Obtaining of Rating Points of Groups of Objects within the Scope of a Qualitative Characteristic

In §5.1—5.2 methods used to obtain rating points of objects within the limits of one and several qualitative characteristics are described. However in practice there are problems of obtaining rating points within the limits of one and several qualitative characteristics for both individual objects, and groups of objects.

Let us consider k groups of objects, for which intensity of manifestation of qualitative characteristic X is estimated within the limits of a verbal scale with levels X_l. Let us denote a relative amount of objects of j-th group referred to level X_l, with a_l^j, and a relative amount of units of all groups referred to level X_l, $l = \overline{1,m}$, with a_l.

Based on a_l, let us construct COSS X with terms X_l, $l = \overline{1,m}$ and membership functions $\mu_l(x) \equiv \left(b_1^l, b_2^l, b_L^l, b_R^l\right)$ of corresponding fuzzy numbers $\tilde{X}_l$. Let us define a fuzzy rating of j-th group of objects within the limits of manifestation of qualitative characteristic X in the form of fuzzy number $\tilde{C}_j$ with membership function

$$\beta_j = a_1^j \otimes \mu_1 \oplus a_2^j \otimes \mu_2 \oplus \ldots \oplus a_m^j \otimes \mu_m, \tag{5.25}$$

or

$$\beta_j \equiv \left(\sum_{l=1}^m a_l^j b_1^l, \sum_{l=1}^m a_l^j b_2^l, \sum_{l=1}^m a_l^j b_L^l, \sum_{l=1}^m a_l^j b_R^l\right). \tag{5.26}$$

Let us define a confidential interval for definite rating z_j, characterizing manifestation of characteristic X for j-th group of objects. With confidence level $\beta_j(z_j) \geq \alpha$, $0 < \alpha < 1$, rating z_j of characteristic X manifestation for j-th group of objects lies within the interval

$$\sum_{l=1}^m a_l^j b_1^l - L^{-1}(\alpha) \sum_{l=1}^m a_l^j b_L^l \leq z_j \leq \sum_{l=1}^m a_l^j b_2^l + R^{-1}(\alpha) \sum_{l=1}^m a_l^j b_R^l.$$

Let us defuzzificate fuzzy numbers with membership functions β_j, $j = \overline{1,k}$, μ_1, μ_m. Let us denote the obtained numbers with C_j, $j = \overline{1,k}$, B_1, B_m. Let us refer the number C_j as pointwise rating of manifestation of qualitative characteristic X for j-th group of objects, $j = \overline{1,k}$.

Let us determine normalized rating E_j of manifestation of qualitative characteristic X for j-th group of objects by the formula

$$E_j = \frac{C_j - B_1}{B_m - B_1}. \tag{5.27}$$

The evaluation E_j is referred to as an average intensity degree of manifestation of characteristic X for j-th group of objects. Range of E_j is a segment $[0, 1]$.

5.4 Obtaining of Rating Points of Groups of Objects within the Limits of Several Qualitative Characteristics

In §5.3 the method of obtaining rating points of groups of objects within the limits of one qualitative characteristic is described; in § 5.4 the method of obtaining rating points of groups of objects within the limits of several qualitative characteristics is offered.

Let us consider k groups of objects for which intensity of manifestations of qualitative characteristics $X_1, ..., X_n$ are estimated. Let us estimate characteristics within the limits of verbal scales with number of levels $k_1, k_2, ..., k_m$. Let us denote a relative number of objects of j-th group, referred to j-th level of a verbal scale while estimating characteristic X_p, $p = \overline{1,m}$, with k_{ij}^p, $i = \overline{1,k}$, $j = \overline{1,k_p}$, $p = \overline{1,m}$, and a relative number of objects of all groups, referred to j-th level of a verbal scale while estimating characteristic X_p, with k_j^p. Based on k_j^p, let us construct m COSS X_p with membership functions of term-sets μ_j^p. Let us compute membership function of the fuzzy rating describing intensity of manifestation of qualitative characteristic X_p for i-th group of objects by the formula

$$\lambda_i^p = k_{i1}^p \otimes \mu_1^p \oplus k_{i2}^p \otimes \mu_2^p \oplus ... \oplus k_{ik_p}^p \otimes \mu_{k_p}^p, \quad i = \overline{1,k}. \tag{5.28}$$

Let us denote weight coefficients (importance coefficients) of characteristics X_p, $\Sigma \omega_p = 1$ with ω_p, $p = \overline{1,m}$. Let us define membership function of the fuzzy rating describing intensity of manifestation of qualitative characteristics $X_1, ..., X_m$ for i-th group of objects, as follows:

$$\lambda_i = \omega_1 \otimes \lambda_i^1 \oplus ... \oplus \omega_m \otimes \lambda_i^m. \tag{5.29}$$

Let us define a confidential interval for obtaining rating z_i characterizing manifestation of qualitative characteristics $X_1, ..., X_m$ for i-th group of objects. Let $\lambda_i \equiv \left(\Delta_1^i, \Delta_2^i, \Delta_L^i, \Delta_R^i \right)$, then with confidence level $\lambda_i(z_i) \geq \alpha$, $0 < \alpha < 1$ rating z_i of manifestation of characteristics $X_1, ..., X_m$ for i-th groups of objects lies in an interval

$$\Delta_1^i - L^{-1}(\alpha)\Delta_L^i \leq z_i \leq \Delta_2^i + R^{-1}(\alpha)\Delta_R^i. \tag{5.30}$$

Let us defuzzificate fuzzy numbers with membership functions

$$\lambda_i, i = \overline{1,k}; \ \eta = \omega_1 \otimes \mu_1^1 \oplus ... \oplus \omega_m \otimes \mu_1^p;$$
$$\delta = \omega_1 \otimes \mu_{k_1}^1 \oplus ... \oplus \omega_m \otimes \mu_{k_p}^p.$$

Let us denote the obtained fuzzy numbers with A_i, $i = \overline{1,k}$, B, C. Number A_i is referred to as a pointwise rating of manifestation of qualitative characteristics $X_1, ..., X_m$ for i-th group of objects, $i = \overline{1,k}$.

Let us compute normalized rating for i-th group of objects by the formula

$$E_i = \frac{A_i - B}{C - B}. \tag{5.31}$$

The evaluation E_i is referred to as an average intensity degree of manifestation of qualitative characteristics $X_1, ..., X_m$ for i-th group of objects. Range of E_i, $i = \overline{1,k}$ is a segment [0, 1].

It is worth mentioning that to operate with various qualitative characteristics, authors use abstract concepts — membership functions of linguistic values of these characteristics. As known, operations for linguistic values are defined on the basis of triangular norms and triangular conorms, and therefore, they are not familiar arithmetical operations.

5.5 Obtaining of Rating Points of Objects within the Limits of Several Quantitative and Qualitative Characteristics

Let us consider a group of N objects, for which quantitative characteristics $X_j, j = \overline{1,l}$ and intensities of manifestation of qualitative characteristics X_v, $v = \overline{l+1,k}$ are estimated. In aggregate, estimated characteristics make essential influence on characteristic Y, for example, success of functioning of objects, being estimated within the limits of the scale: Y_1 = "extremely unsuccessful", Y_2 = "unsuccessful", Y_3 = "mean successful", Y_4 = "rather successful", Y_5 = "extremely successful".

Ranges of values of quantitative characteristics X_j can be non-enumerable sets of points R_j of the real values line.

Using expert information we construct COSS with names X_j. If growth of characteristic X_j is accompanied with growth of characteristic Y, then «very small value of characteristic X_j», "small value of characteristic X_j", "average value of characteristic X_j", "large value of characteristic X_j", "very large value of characteristic X_j" are COSS terms, and $\mu_{ij}(x)$, $i = \overline{1,5}$ are their membership functions. If growth of characteristic X_j is accompanied by decreasing of characteristic Y, then "very great value of characteristic X_j", "great value of characteristic X_j"," average value of characteristic X_j", "small value of characteristic X_j", "very small value of characteristic X_j" are COSS terms, and $\mu_{ij}(x)$, $i = \overline{1,5}$ are their membership functions.

Let us denote values of characteristics X_j for n-th object, $n = \overline{1,N}$ with x_j^n, and degrees of a membership of these values to COSS terms named X_j – with $\mu_{ij}(x_j^n)$, $i = \overline{1,5}$, $j = \overline{1,l}$, $n = \overline{1,N}$.

Let X_{lv} be levels of the verbal scales applied to an evaluation of characteristics X_v, $v = \overline{l+1,k}$. Levels are arranged in ascending order of manifestation intensity of relevant characteristic, if its growth is accompanied with Y growth, and in decreasing order if its growth is accompanied with Y decrease.

Let us construct $k - l$ COSS's named X_v, having related term-sets X_{lv}, and membership functions $\mu_{lv}(x)$. $U = [0,1]$ is selected as universal COSS sets. Fuzzy numbers $\tilde{X}_{lv}$, $l = \overline{1,m_v}$, $v = \overline{l+1,k}$ or their membership functions $\mu_{lv}(x)$ are referred to as evaluations of objects. Let us denote an evaluation of n-th object within the limits of characteristic X_v with $\tilde{X}_v^n$ and $\mu_v^n(x) \equiv (a_{v1}^n, a_{v2}^n, a_{vL}^n, a_{vR}^n)$. Fuzzy number $\tilde{X}_v^n$ with membership function $\mu_v^n(x)$ is equal to one of fuzzy numbers $\tilde{X}_{lv}$, $l = \overline{1,m_v}$, $v = \overline{l+1,k}$.

Let us denote with $\delta_i(x_v^n)$, $n = \overline{1,N}$, $v = \overline{l+1,k}$, $i = \overline{1,5}$ the function which equal to '1' if an evaluation of n-th object within the limits of characteristic X_v is fuzzy number $\tilde{X}_{iv}$, $i = \overline{1,5}$, and equal to zero if an evaluation of n-th object within the limits of characteristic X_v is fuzzy number $\tilde{X}_{pv}$, $p = \overline{1,5}$, $p \neq i$.

Let us denote with ω_j, $j = \overline{1,k}$, $\sum_{j=1}^{k} \omega_j = 1$ weight coefficients of estimated characteristics. Calculate the following coefficients:

$$\lambda_i^n = \sum_{j=1}^{l} \omega_j \mu_{ij}\left(x_j^n\right) + \sum_{v=l+1}^{k} \omega_v \delta_i\left(x_v^n\right), \ i = \overline{1,5}, \ n = \overline{1,N}. \tag{5.32}$$

Let us compute the sum of these coefficients, using characteristics of COSS and definition of functions $\delta_i\left(x_v^n\right)$, $n = \overline{1,N}$, $v = \overline{l+1,k}$, $i = \overline{1,5}$.

$$\begin{aligned}
\sum_{i=1}^{5} \lambda_i^n &= \omega_1\left[\mu_{11}\left(x_1^n\right) + \mu_{21}\left(x_1^n\right) + \ldots + \mu_{51}\left(x_1^n\right)\right] + \ldots \\
&+ \omega_l\left[\mu_{1l}\left(x_l^n\right) + \mu_{2l}\left(x_l^n\right) + \ldots + \mu_{5l}\left(x_l^n\right)\right] + \\
&+ \omega_{l+1}\left[\delta_1\left(x_{l+1}^n\right) + \delta_2\left(x_{l+1}^n\right) + \ldots + \delta_5\left(x_{l+1}^n\right)\right] + \ldots \\
&+ \omega_k\left[\delta_1\left(x_k^n\right) + \delta_2\left(x_k^n\right) + \ldots + \delta_5\left(x_k^n\right)\right] = \sum_{i=1}^{5} \omega_j = 1.
\end{aligned}$$

Based on the above, we may consider coefficients λ_i^n, $i = \overline{1,5}$, $n = \overline{1,N}$ as weight coefficients of terms of characteristic Y for n-th object, $n = \overline{1,N}$. A fuzzy rating of n-th object within the limits of characteristics X_j, $j = \overline{1,k}$ is determined as fuzzy number

$$\tilde{A}_n = \lambda_1^n \otimes \tilde{Y}_1 \oplus \ldots \oplus \lambda_5^n \otimes \tilde{Y}_5 \tag{5.33}$$

with membership function

$$\mu_n(x) \equiv \left(\sum_{i=1}^{5} \lambda_i^n a_{i1}, \ \sum_{i=1}^{5} \lambda_i^n a_{j2}, \ \sum_{i=1}^{5} \lambda_i^n a_{jL}, \ \sum_{i=1}^{5} \lambda_i^n a_{iR} \right),$$

where $\tilde{Y}_i \equiv \left(a_{i1}, a_{i2}, a_{iL}, a_{iR}\right)$.

Let us define a confidential interval for obtaining definite rating y_n. If confidence level is $\mu_n(y_n) \geq \alpha$, $0 < \alpha < 1$, rating y_n of n-th object lies within the interval

$$\sum_{i=1}^{5} \lambda_i^n a_{j1} - (1-\alpha)\sum_{i=1}^{5} \lambda_i^n a_{jL} \leq y_n \leq \sum_{i=1}^{5} \lambda_i^n a_{j2} + (1-\alpha)\sum_{i=1}^{5} \lambda_i^n a_{jR}.$$

Let us defuzzificate fuzzy number $\tilde{A}_n$ using the gravity method; let us denote the obtained definite number with A_n.

To recognize success of objects' functioning it is necessary to identify fuzzy number having membership function $\mu_n(x)$ with one of COSS terms named Y (with one of fuzzy numbers $\tilde{Y}_i$, $i = \overline{1,5}$ with membership functions $\mu_i(x)$, $i = \overline{1,5}$). For this purpose let us calculate identification indexes:

$$\beta_n^i = \frac{\int_0^1 \min[\mu_i(x), \mu_n(x)]dx}{\int_0^1 \max[\mu_i(x), \mu_n(x)]dx}, \quad i = \overline{1,5}, \quad n = \overline{1,N}. \tag{5.34}$$

If $\beta_n^p = \max_i \beta_n^i$, the state of n-th object is defined by p-th by level of scale Y_1 = "extremely unsuccessful", Y_2 = "unsuccessful", Y_3 = "mean successful", Y_4 = "rather successful", Y_5 = "extremely successful", $p = \overline{1,5}$.

Let us denote ratings of n-th object for phases 1 and 2 with A_n^1, A_n^2 accordingly. Depending on ratio of A_n^1 to A_n^2, it is possible to draw the following conclusions: if $A_n^1 > A_n^2$, the state of n-th object is worsened; if $A_n^1 < A_n^2$ the state of n-th object is improved; if $A_n^1 = A_n^2$, the state of n-th object is unchanged.

5.6 Examples of Practical Application of the Methods Developed

Example 5.1. Obtaining of rating points of trainees through their academic achievements during a semester. Rating systems of knowledge evaluation are widely applied in educational process and play an essential role in the education quality control aspects. These systems allow, at any grade level, obtaining accessible and opportune information in the form of some integral index used for making some administrative decisions. Rating systems of knowledge evaluation are purposed for lowering the subjectivity between teachers and trainees, and also to eliminate other (probably latent) coefficients hindering to objectively evaluate level of training.

Let us consider a problem of obtaining rating points of knowledge of students through their academic achievements during a semester. Calculation-graphic tasks (CGT) and tests (T) in linear algebra, analytical geometry and following sections of mathematical analysis: limits, derivatives and indefinite integrals, were estimated with marks from zero to ten points. In addition, independent work and class-work were evaluated with marks from zero to ten points. Let us assume that all types of educational activities have equal weight coefficients. Results of an evaluation of knowledge of ten trainees are shown in Table 5.1.

Using the method described in §2.2, COSS "knowledge of students studying higher mathematics" is constructed. Data necessary for model-building are obtained on the base from information available in the previous experience of a teacher. Membership functions $\mu_i(x)$, $l = \overline{1,4}$ of term-sets "F", "C", "B", "A", accordingly, are membership functions of T numbers or normal triangular numbers and have parameters

$$\left.\begin{array}{ll} \mu_1(x) = (0; 0,1; 0; 0,2); & \mu_2(x) = (0,3; 0,5; 0,2; 0,2); \\ \mu_3(x) = (0,7; 0,2; 0,2); & \mu_4(x) = (0,9; 1; 0,2; 0). \end{array}\right\} \tag{5.35}$$

Table 5.1 Results of knowledge evaluation of trainees during a semester

Subject	1	2	3	4	5	6	7	8	9	10
Linear algebra, CGT	5	7	2	8	9	3	6	0	1	9
Linear algebra, T	6	7	1	7	8	4	7	2	1	10
Analytical geometry, CGT	7	8	4	0	9	7	6	4	1	9
Analytical geometry, T	6	7	6	8	8	8	7	5	1	10
Limits, CGT	5	6	6	9	9	7	7	6	0	9
Limits, T	6	6	7	9	9	8	8	6	1	9
Derivatives, CGT	7	7	7	8	9	8	7	7	1	9
Derivatives, T	9	9	10	9	9	9	10	10	1	9
Indefinite integral, CGT	2	3	4	3	5	5	8	8	1	10
Indefinite integral, T	4	4	5	4	6	6	9	9	0	10
Independent work	6	6	7	6	8	9	10	10	1	10
Classwork	7	8	4	4	5	8	9	9	1	9

Table 5.2 Rating points of knowledge and traditional marks of knowledge evaluation of trainees in higher mathematics

Number of the trainee	E_n	Traditional marks of knowledge evaluation
1	0,506	"C"
2	0,611	"C"
3	0,475	"C"
4	0,672	"C"
5	0,744	"B"
6	0,661	"C"
7	0,735	"B"
8	0,568	"C"
9	0	"F"
10	1	"A"

Using the method described in this paragraph, we obtain the results summarized in Table 5.2.

Example 5.2. Obtaining of experts rating points. Let us consider an issue of staff structure [176—179] which consists in obtaining rating points of employees and assigning one of four qualification levels X_1, X_2, X_3, X_4 to each employee. Levels are arranged in ascending order of their relevant ratings. The questionnaire offered to employees consists of five sections related to educational level, scientific degree, age, and work experience and language qualifications.

X_1 — an educational level. Answers: secondary education — 0 points, specialized secondary education — 1 point, incomplete higher education — 2 points, bachelor's degree — 3 points, higher education (specialist or master's degree) — 4 points.

X_2 — scientific degree. Answers: none — 0 points, Candidate of Science (PhD) — 1 point, Doctor of Science (Grand PhD) — 2 points.

X_3 — age. Answers: more than 45 years — 0 points, 35 - 45 years — 1 point, under 35 years — 2 points.

X_4 — employment experience in the considered occupation. Answers: none — 0 point, less than 5 years — 1 point, 5 years to 10 years — 2 points, 10 to 15 years — 3 points, more than 15 years — 4 points.

X_5 — linguistic skill. Answers: none - 0 points; I read and I translate texts with dictionary — 1 point; I make myself understood at basic level — 2 points; fluent language— 3 points. The management of human resource department considered the following percentage ratio of employees in frameworks of each level as ideal: X_1 — 20 %, X_2 — 30 %, X_3 — 40 %, X_4 — 10 %. Based on the above, indexes $a_1 = 0,2$, $a_2 = 0,3$, $a_3 = 0,4$, $a_4 = 0,1$ are obtained, and using the method described in §2.1, membership functions μ_i, $i = \overline{1,4}$ for corresponding terms X_i, $i = \overline{1,4}$, are constructed, they are membership functions of T -numbers and have parameters:

$$\mu_1(x) = (0; 0,1; 0; 0,2), \ \mu_2(x) = (0,3; 0,35; 0,2; 0,3),$$
$$\mu_3(x) = (0,65; 0,85; 0,3; 0,1), \ \mu_4(x) = (0,95; 1; 0,1; 0).$$

Weight coefficients for sections of the questionnaire are allocated as following: $\omega_1 = 0,4$, $\omega_2 = 0,1$, $\omega_3 = 0,2$, $\omega_4 = 0,2$, $\omega_5 = 0,1$. Using the methods described in §5.5, we obtained the results shown in Table 5.3.

From Table 5.3. one can see that, for example, the respondents No.6 and No. 12 have identical number of points

$$\sum_{j=1}^{5} y_j^n = 6.$$

Table 5.3 Results of questioning of fifteen employees

No.	y_1^n	y_2^n	y_3^n	y_4^n	y_5^n	s_1^n	s_2^n	s_3^n	s_4^n	s_5^n	$\sum y_j^n$	E_n	Level
0	0	2	1	0	0	0	1	1/4	0	3	0,245	0	X_2
1	0	2	2	0	1/4	0	1	1/2	0	5	0,388	1	X_2
2	0	1	3	1	1/2	0	1/2	3/4	1/3	7	0,465	2	X_2
4	1	2	4	3	1	1/2	1	1	1	14	0,949	4	X_4

Table 5.3 (*continued*)

1	0	0	1	1	1/4	0	0	1/4	1/3	3	0,428	1	X_2
1	0	2	2	1	1/4	0	1	1/2	1/3	6	0,417	1	X_2
3	0	2	3	1	3/4	0	1	3/4	1/3	9	0,647	3	X_3
0	0	1	0	0	0	0	1/2	0	0	1	0,200	0	X_1
0	0	2	1	1	0	0	1	1/4	1	4	0,399	0	X_2
4	2	2	4	3	1	1	1	1	1	15	1	4	X_4
2	2	0	4	2	1/2	1	0	1	2/3	10	0,567	2	X_2
1	0	0	4	1	1/4	0	0	1	1/3	6	0,319	1	X_2
0	0	1	4	1	0	0	1/2	1	1/3	6	0,329	0	X_2
2	2	0	4	2	1/2	1	0	1	2/3	10	0,567	2	X_2
0	0	0	0	0	0	0	0	0	0	0	0	0	X_1

If we use of widely applied method of rating computation by the formula

$$x^n = \sum_{j=1}^{5} y_j^n \omega_j,$$

then it appears that the respondents No.6 and No.12 have the identical rating points equal to 1.3 in this case, too.

The developed method of obtaining rating points and assignment of qualification levels allows obtaining more information on qualification of employees in comparison with a traditional method within the limits of the same information. The respondents No.6 and No.12 are assigned to level X_2, however, they have different rating points. Fig. 5.1 shows the functional model of evaluation of correspondence of specialists' training level to requirements of their professional activity sphere.

Example 5.3. Professional selection of graduates [180—183]. One of the major problems of professional selection is the problem of detection of candidates which could master a particular specialty within target dates and further effectively fulfill their professional duties, by virtue of their educational level within the limits of subject matters, and also developmental level of personal qualities and cognitive psycho-physiological processes.

The functional model of multi-criteria professional selection of graduates accounting for fuzzy preferences is shown in Fig. 5.2.

Let us consider academic performance indexes, cognitive psycho-physiological and personality traits of graduates of a certain specialty. The general educational level is divided into fundamental sciences education (FE), general professional training (GPT) and specialized education (SE).

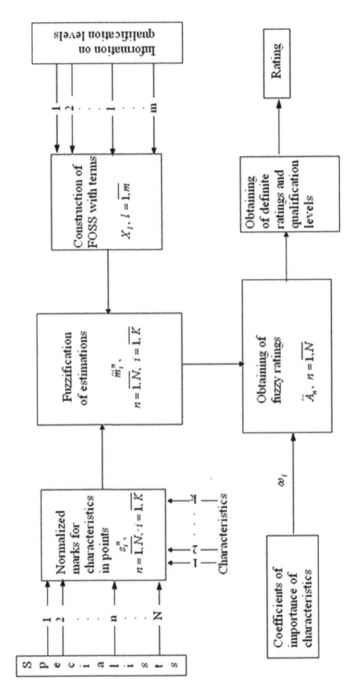

Fig. 5.1 Functional model of evaluation of specialists

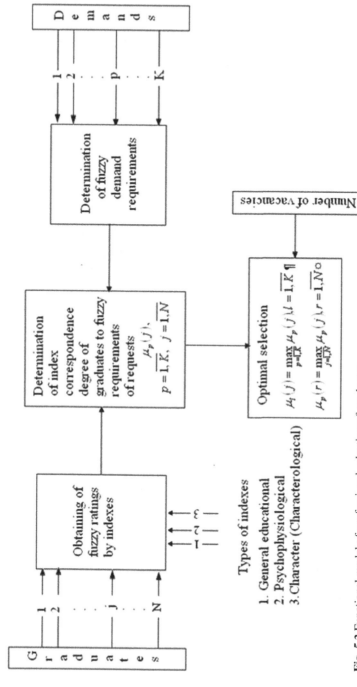

Fig. 5.2 Functional model of professional selection of graduates

Table 5.4 Academic achievements of graduates within the limits of NSE, SE, and LAE

Discipline (subject)	"C" (satisfactory)	"B" (good)	"A" (excellent)
within the limits of NSE			
Higher mathematics	38	32	30
Physics	41	37	22
within the limits of SE			
Political science	43	19	38
Jurisprudence	31	35	34
within the limits of LAE			
Psychology and Pedagogy	11	39	50
Philosophy	24	42	34
Foreign language	33	33	34
Russian history	16	52	32

Level of fundamental sciences education of graduates was defined on the basis of the analysis of their progress for all period of studies in high school for series of general science disciplines: higher mathematics, physics, Russian history, foreign language, psychology and pedagogy, philosophy, jurisprudence, political science. For more detailed analysis, FE is divided into natural sciences education (NSE), social and economic sciences education (SESE) and liberal arts education (humanitarian training) (LAE). The considered data of graduates' progress for each of FE series are summarized in Table 5.4 – 5.5.

According to data from Table 5.4 eight (as per number of disciplines) COSS's are constructed with "knowledge of graduates" titles. All COSS's have terms "C","B", "A" with membership functions of T-numbers or normal triangular numbers. Being constructing, fuzzy numbers corresponding to terms "C", "B", "A" are put in correspondence to the marks of graduates. Applying the method considered in §5.2, ratings of graduates are obtained within the limits of NSE, SESE, LAE and FE. Weight coefficients of all disciplines are taken equal. For analysis of progress of five graduates, their average indexes of progress have compared (Table 5.5) to the rating points obtained using the method described in §5.2.

Normalized rating points in Table 5.5 are quantity indexes of average intensity degree of manifestation of knowledge by the graduates within the limits of NSE, SESE, LAE and FE. From data of Table 5.5 one can understand that the results obtained on the basis of fuzzy rating points considerably enlarge the results obtained on the basis of average evaluations. For example, graduates No.3 and No.4 have identical average evaluations within the limits of NSE, but different normalized ratings.

Table 5.5 Progress of five graduates within the limits of FE and their normalized rating points

Discipline	Number of the graduate				
	1	2	3	4	5
Progress level within the limits of FE					
Higher mathematics	4	5	3	4	5
Physics	4	4	4	3	5
Political science	4	3	5	5	4
Jurisprudence	5	4	5	5	4
Psychology and Pedagogy	4	3	4	4	5
Philosophy	5	4	5	5	4
Foreign language	4	3	4	4	5
Russian history	4	4	5	5	5
NSE	4	4,5	3,5	3,5	5
SE	4,5	3,5	5	5	4
LAE	4,25	3,5	4,5	4,5	4,75
FE	4,25	3,83	4,33	4,33	4,58
Normalized ratings within the limits of					
NSE, SESE, LAE and FE					
NSE	0,541	0,770	0,288	0,266	1
SE	0,768	0,263	1	1	0,507
LAE	0,759	0,425	0,870	0,870	0,917
FE	0,617	0,379	0,682	0,676	0,812

GPT and SE levels of graduates were defined on the basis of the analysis of their progress for all period of studies in high school in the corresponding disciplines. For each specialty these disciplines were selected according to curriculums and curricula. The list of disciplines within the limits of all educational courses can be composed based on the analysis of correlation of testimonials from employers about professional work of graduates and evaluations of academic achievements of graduates for all period of studies. According to progress of graduates within the limits of GPT and SE data, the average evaluation marks of graduates (Table 5.6 and 5.7) are calculated.

Applying the method described in § 5.2, normalized ratings of graduates within the limits of GPT and SE are obtained.

Table 5.6 Progress of all graduates within the limits of GPT and the SE, in %, and five graduates separately

Discipline (subject)	"C" (satisfactory)	"B" (good)	"A" (excellent)
	within the limits of GPT, %		
1	34	31	35
2	10	23	67
3	16	45	39
4	9	41	50
5	32	36	32
	within the limits of SE, %		
6	25	32	43
7	21	32	47
8	9	56	35

	Progress of five graduates				
	1	2	3	4	5
	within the limits of GPT, %				
1	4	3	5	4	4
2	4	4	4	5	5
3	3	5	3	4	4
4	4	4	4	4	4
5	5	5	4	5	5
average evaluation mark	4	4,2	4	4,4	4,4
	within the limits of SE, %				
6	4	5	4	5	5
7	3	4	3	4	4
8	4	3	4	5	4
average evaluation mark	3,66	4	3,66	4,66	4,33

Table 5.7 Normalized rating points of five graduates within the limits of GPT and SE

Rating	Number of the graduate				
	1	2	3	4	5
within the limits of GPT	0,425	0,542	0,422	0,649	0,649
within the limits of SE	0,299	0,439	0,299	0,810	0,601

Weight coefficients of all disciplines occurred to be equal (Table 5.7).

Assuming weight coefficients of all considered subjects equal, we obtained fuzzy rating points and normalized ratings of five graduates within the limits of the general educational level. Parameters of fuzzy rating points are given in the columns of Table 5.8.

The evaluation of cognitive psychophysiological characteristics (general intellectual abilities) of graduates was spent by means of the original instant diagnosis mentality test developed in Scientific research institute of professional selection. The test is related to the category of general intellectual ability tests. Being tested, a person is offered a series of tasks selected so that the adequate evaluation of the major intellectual functions are ensured. The mentality test gives the opportunity to define a degree of development of such major components of intelligence, as thinking logicality, adequate perception, mental speed, spatial perception, literacy. All components are evaluated within the limits of a verbally-numerical scale: "low" — 2 points, "mean" — 3 points, "high" — 4 points, "very high" — 5 points (Table 5.9 and Table 5.10).

Table 5.8 Normalized and fuzzy ratings of the general educational level of graduates on the basis of their progress

| Rating | Number of the graduate | | | | |
	1	2	3	4	5
Normalized rating	0,497	0,441	0,527	0,694	0,722
Fuzzy rating	0,426	0,487	0,519	0,674	0,709
	0,532	0,568	0,686	0,747	0,839
	0,113	0,109	0,106	0,088	0,132
	0,097	0,058	0,113	0,128	0,145

Table 5.9 Results of graduates testing

Item	"Low"	"Mean"	"High"	"Very high"
Thinking logicality	20	21	37	22
Adequate perception	2	4	42	52
Mental speed				
Spatial perception	12	8	45	35
Literacy	42	30	15	13
Thinking logicality	18	15	43	24

Table 5.10 Evaluation of cognitive psychophysiological characteristics of five graduates

Item	No. of the graduate				
	1	2	3	4	5
Thinking logicality	3	3	3	3	4
Adequate perception	4	4	5	4	5
Mental speed					
Spatial perception	2	3	2	3	3
Literacy	3	3	2	4	3
Mean marks	4	3	4	5	5
Thinking logicality	3,2	3,2	3,2	3,8	4

Applying the method described in § 5.2, we obtained fuzzy and normalized rating points of cognitive psychophysiological characteristics of graduates. Weight coefficients of their components was assumed being equal. Parameters of fuzzy rating points are summarized in Table 5.11.

Besides, academician facilitators of study groups performed expert evaluations of developmental level of personal traits of graduates. They evaluated graduates in terms of the following qualities: public activity, discipline and promptitude in obeying, diligence, consistency and self-control, organizing abilities, group opinion leader, purposefulness, within the limits of a verbally-numerical scale: "low" — 2 points, "mean" — 3 points, "high" — 4 points, "very high" — 5 points. Table 5.12 shows data resulted from the evaluation by academician facilitators of study groups for the personal qualities of graduates, and Table 5.13 — evaluations of personal qualities of five graduates whose progress data is presented above.

Table 5.11 Normalized and fuzzy rating points of cognitive psychophysiological characteristics of five graduates

Rating	No. of the graduate				
	1	2	3	4	5
Normalized	0,343	0,290	0,367	0,517	0,648
Fuzzy	0,312	0,271	0,341	0,498	0,613
	0,376	0,342	0,396	0,539	0,698
	0,114	0,094	0,029	0,126	0,115
	0,204	0,126	0,095	0,103	0,118

Table 5.12 Results of the evaluation of personal qualities of graduates made by academician facilitators of study groups, %

Character traits	"Low"	"Mean"	"High"	"Very high"
public activity	14	66	16	4
discipline and promptitude in obeying	6	16	36	42
diligence	18	30	30	22
consistency and self-control	16	44	22	18
organizing abilities	46	28	18	8
group opinion leader	14	38	42	6
purposefulness	24	42	20	14

Table 5.13 Evaluation of personal qualities of five graduates

Qualities	No. of the graduate				
	1	2	3	4	5
public activity	4	4	3	3	4
discipline and promptitude in obeying	4	5	4	5	5
diligence	3	2	3	4	4
consistency and self-control	4	3	5	5	2
organizing abilities	2	2	3	5	4
group opinion leader	4	3	4	3	3
purposefulness	4	4	4	5	5
Average marks	3,57	3,27	3,71	4	3,85

While obtaining fuzzy rating points and normalized ratings of personal qualities of graduates, weight coefficients of all components were considered be equal. Parameters of fuzzy ratings are summarized in Table 5.14.

Table 5.14 Normalized and fuzzy rating points of personal qualities of five graduates

Rating	No. of the graduate				
	1	2	3	4	5
Normalized	0,585	0,484	0,610	0,754	0,657
Fuzzy	0,574	0,468	0,602	0,732	0,635
	0,598	0,488	0,625	0,768	0,687
	0,104	0,112	0,109	0,096	0,116
	0,136	0,142	0,134	0,094	0,211

Selection of graduates was carried out within the limits of four formulated fuzzy preferences:

X_1 — progress indexes are very important, indexes of cognitive psychophysiological characteristics are very important and indexes of personal qualities are absolutely unimportant;

X_2 — Progress indexes are not so important, indexes of cognitive psychophysiological characteristics are important enough, and indexes of personal qualities are very important;

X_3 — Progress indexes are rather unimportant, indexes of cognitive psychophysiological characteristics are important, and indexes of personal qualities are rather unimportant;

X_4 — progress indexes are absolutely unimportant, indexes of cognitive psychophysiological characteristics are important enough, and indexes of personal qualities are important,

The memberships functions $\mu_l(x)$, $l = \overline{1,6}$ constructed in (49) for corresponding linguistic terms "absolutely unimportant", "rather unimportant", "not so important", "important enough", "important", "very important" of relative importance coefficients for criteria:

$$\mu_1(x) \equiv (0; 0; 0,2); \quad \mu_2(x) \equiv (0,2; 0,2; 0,2);$$
$$\mu_3(x) \equiv (0,4; 0,2; 0,2); \quad \mu_4(x) \equiv (0,6; 0,2; 0,2);$$
$$\mu_5(x) \equiv (0,8; 0,2; 0,2); \quad \mu_6(x) \equiv (1; 0,2; 0).$$

were used.

Membership functions of fuzzy rating progress evaluations, psychophysiological characteristics and personal qualities of five graduates are denoted, accordingly, with $\eta_j^1(x)$, $\eta_j^2(x)$, $\eta_j^3(x)$, $j = \overline{1,5}$. Using the method of ranking [184], membership functions $\mu_{R_j^i}(x)$, $j = \overline{1,5}$, $i = \overline{1,4}$ of fuzzy evaluations of five graduates $\overline{R}_j^i$, $j = \overline{1,5}$, $i = \overline{1,4}$ within the limits of four formulated fuzzy preferences were described as follows:

$$\mu_{R_j^1}(x) = \mu_6(x) \otimes \eta_j^1(x) \oplus \mu_6(x) \otimes \eta_j^2(x) \oplus \mu_1(x) \otimes \eta_j^3(x);$$
$$\mu_{R_j^2}(x) = \mu_3(x) \otimes \eta_j^1(x) \oplus \mu_4(x) \otimes \eta_j^2(x) \oplus \mu_6(x) \otimes \eta_j^3(x);$$
$$\mu_{R_j^3}(x) = \mu_2(x) \otimes \eta_j^1(x) \oplus \mu_5(x) \otimes \eta_j^2(x) \oplus \mu_2(x) \otimes \eta_j^3(x);$$
$$\mu_{R_j^4}(x) = \mu_1(x) \otimes \eta_j^1(x) \oplus \mu_4(x) \otimes \eta_j^2(x) \oplus \mu_5(x) \otimes \eta_j^3(x), \quad j = \overline{1,5}.$$

Results of researches for five graduates were carried out on the basis of $\tilde{R}_j^i$, $j = \overline{1,5}$, $i = \overline{1,4}$. Fuzzy sets I^i, $i = \overline{1,4}$ specified on set of indexes $\{1, 2, 3, 4, 5\}$

were determined. Values of membership functions $\mu_i(j)$ of these sets are interpreted as characteristics of domination degree of j-th graduate within the limits of fuzzy preferences X_i, $i = \overline{1,5}$, $j = \overline{1,4}$ or characteristics of to what extent the j-th graduate is considered the best within the limits of fuzzy preferences X_i, $i = \overline{1,5}$, $j = \overline{1,4}$; k-th the graduate was considered the best within the limits of fuzzy preference X_i, with the characteristic equal to '1', if $\sup x : \mu_{R_j^i}(x) = 1$ belonged to $\tilde{R}_k^i(x)$. Values $\mu_i(j)$ with $j \neq k$ are calculated as follows:

$$\mu_i(j) = \max_x \min\left[\mu_{R_j^i}(x), \mu_{R_k^i}(x)\right]$$

The calculated characteristics $\mu_i(j)$, $j = \overline{1,5}$, $i = \overline{1,4}$ are summarized in Table 5.15.

Based on the analysis carried out, graduate No. 1 is recommended for job placement option X_4, graduate No.2 — job placement option X_2, graduate No. 3 — job placement option X_2, graduate No. 4 - job placement options X_2, X_3, X_4, and graduate No.5 is recommended for job placement options X_1, X_2, X_4.

Table 5.15 Characteristics of domination degree of graduates under fuzzy preferences

No. of the graduate	Domination degree			
	1	2	3	4
1	0,53	0,61	0,77	0,79
2	0,45	0,76	0,68	0,66
3	0,73	0,85	0,82	0,83
4	0,96	1	1	1
5	1	1	0,98	1

Example 5.4. [185-186] Obtaining of rating points of plant species in the conditions of big cities. Let us consider data from the example 3.5 and obtain rating points of a state of considered plant species in plantings of the Boulevard Ring avenue in Moscow, and for comparison – rating points of the state of the same plant species in the parkways far from the centre of Moscow (5022 plants were inspected).

Apparently, because of different ecological conditions in the city centre and in suburbs, it was to be expected different rating points for the same plant species that is explained through their different adaptation to difficult ecological conditions within the big city.

The relative number of objects of i-th specie in the plantings of the Boulevard Ring avenue referred by experts to l-th level of a verbal scale is denoted with a_{il}^1, $i = \overline{1,17}$, $l = \overline{1,7}$. The relative quantity of plants of i-th specie in plantings of other parkways referred to l-th level of a verbal scale is denoted with a_{il}^2, $i = \overline{1,17}$, $l = \overline{1,7}$, the relative quantity of plants of seventeen species in the plantings of the Boulevard Ring avenue referred to l-th level of a verbal scale is denoted with a_l^1, $l = \overline{1,7}$, and the relative quantity of plants of seventeen species in the plantings of other parkways referred to l-th level of a verbal scale is denoted with a_l^2, $l = \overline{1,7}$.

In Table 3.15 data a_{il}^1, $i = \overline{1,17}$, $l = \overline{1,7}$ are shown, in Table 5.16 data a_l^1, $l = \overline{1,7}$ are shown, in Table 5.17 data a_l^2, $l = \overline{1,7}$ are shown, and in Table 5.18 data a_{il}^2, $i = \overline{1,17}$, $l = \overline{1,7}$ are shown.

According to Table 3.15 data, COSS= "condition of plantings of the Boulevard Ring avenue" is constructed, and according to Table 5.17 data, COSS= "condition of plantings in other parkways" is constructed. Fuzzy numbers $\tilde{X}_l^1$, $\tilde{X}_l^2$, $l = \overline{1,7}$ corresponding to terms of COSS "condition of plantings of Boulevard Ring" and "condition of plantings in other parkways", are presented in the form of T-numbers or normal triangular numbers. Parameters of membership functions μ_l^1, $l = \overline{1,7}$ of terms of COSS = "condition of plantings on the Boulevard Ring avenue" are given in Table 5.19, and parameters of membership functions for terms COSS = "condition of plantings in other parkways" are summarized in Table 5.20.

Membership functions of the fuzzy rating points describing conditions of seventeen plant species in plantings of the Boulevard Ring avenue are defined by the formula

$$\lambda_i^1 = a_{i1}^1 \otimes \mu_1^1 \oplus a_{i2}^1 \otimes \mu_2^1 \oplus ... \oplus a_{i7}^1 \otimes \mu_7^1, \ i = \overline{1,17}. \tag{5.36}$$

Table 5.16 Relative numbers of plants of the Boulevard Ring avenue within the limits of levels of a verbal scale

Place of growth of a plant	a_1^1	a_2^1	a_3^1	a_4^1	a_5^1	a_6^1	a_7^1
The Boulevard Ring avenue	0,009	0,010	0,030	0,130	0,289	0,437	0,095

Table 5.17 Relative numbers of plants in other parkways within the limits of levels of a verbal scale

Place of growth of a plant	a_1^2	a_2^2	a_3^2	a_4^2	a_5^2	a_6^2	a_7^2
Other parkways	0,004	0,049	0,058	0,269	0,430	0,149	0,042

Table 5.18 Relative results of condition evaluation for plant species in other parkways

№	Name of a specie	a_{i1}^2	a_{i2}^2	a_{i3}^2	a_{i4}^2	a_{i5}^2	a_{i6}^2	a_{i7}^2
1	European white birch	0,009	0,032	0,060	0,100	0,660	0,070	0,069
2	European hawthorn	0,000	0,009	0,180	0,290	0,450	0,030	0,041
3	European white elm	0,000	0,020	0,048	0,510	0,296	0,060	0,066
4	Witch elm	0,022	0,000	0,043	0,365	0,365	0,182	0,023
5	Single-seed hawthorn	0,000	0,054	0,055	0,363	0,253	0,154	0,121
6	Cotoneaster	0,000	0,000	0,050	0,230	0,670	0,050	0,000
7	Norway maple	0,006	0,054	0,030	0,087	0,468	0,235	0,120
8	Tatarian maple	0,000	0,000	0,000	0,139	0,805	0,028	0,028
9	Canadian maple	0,000	0,026	0,068	0,273	0,462	0,120	0,051
10	Large-leaved linden	0,000	0,008	0,061	0,220	0,365	0,262	0,084
11	Little-leaved linden	0,000	0,027	0,073	0,304	0,419	0,145	0,032
12	Hungarian lilac	0,014	0,021	0,021	0,119	0,594	0,203	0,028
13	Common lilac	0,010	0,019	0,060	0,235	0,413	0,255	0,008
14	Cottonwood	0,000	0,019	0,029	0,466	0,310	0,049	0,127
15	Rough-bark poplar	0,001	0,160	0,037	0,303	0,417	0,070	0,012
16	European ash	0,000	0,079	0,048	0,190	0,365	0,270	0,048
17	Black ash	0,003	0,062	0,074	0,356	0,411	0,067	0,027

Table 5.19 Parameters of membership functions of term-sets of COSS "condition of plantings in the Boulevard Ring avenue"

Function	Parameter of the function			
μ_1^1	0,000	0,005	0,000	0,009
μ_2^1	0,014	0,014	0,009	0,010
μ_3^1	0,024	0,034	0,010	0,030
μ_4^1	0,064	0,114	0,030	0,130
μ_5^1	0,244	0,324	0,130	0,289
μ_6^1	0,613	0,858	0,289	0,095
μ_7^1	0,953	1,000	0,095	0,000

Table 5.20 Parameters of membership function of term-sets of COSS "condition of plantings in other parkways"

Function	Parameter of the function			
μ_1^2	0,000	0,002	0,000	0,004
μ_2^2	0,006	0,029	0,004	0,049
μ_3^2	0,078	0,082	0,049	0,058
μ_4^2	0,140	0,246	0,058	0,269
μ_5^2	0,515	0,736	0,269	0,149
μ_6^2	0,885	0,938	0,149	0,042
μ_7^2	0,980	1,000	0,042	0,000

Numbers a_{il}^1, $i = \overline{1,17}$, $l = \overline{1,7}$ are shown in Table 3.15 of the example 3.5.

Membership functions of the fuzzy rating points describing conditions of seventeen plant species in plantings of other parkways, are computed by the formula

$$\lambda_i^2 = a_{i1}^2 \otimes \mu_1^2 \oplus a_{i2}^2 \otimes \mu_2^2 \oplus ... \oplus a_{i7}^2 \otimes \mu_7^2, \; i = \overline{1,17},$$

where $\otimes$, $\oplus$ —generalized operations of multiplication and summation, accordingly.

Fuzzy numbers λ_i^1, λ_i^2, $i = \overline{1,17}$, $\tilde{X}_1^1$, $\tilde{X}_7^1$, $\tilde{X}_1^2$, $\tilde{X}_7^2$ are defuzzificated by the gravity method. The obtained numbers are denoted accordingly with C_i^1, C_i^2, $i = \overline{1,17}$, A_1^1, A_7^1, A_1^2, A_7^2. Normalized ratings of seventeen plant species in plantings of the Boulevard Ring avenue and in plantings of other parkways are computed by the formula

$$E_i^p = \frac{C_i^p - A_1^p}{A_7^p - A_1^p}, \; p = \overline{1,2}, \; i = \overline{1,17}.$$

Based on the obtained rating points, each plant specie is assigned to the state rating in accordance with the approach saying: the higher rating points are, higher the rating is. The obtained results are summarized in Table 5.21 and Table 5.22.

The analysis completed allowed estimating the stability of individual species of woody plants and brushwood under conditions of intensive human intervention.

Table 5.21 Fuzzy rating points, rating points and a rating of species of woody plants and brushwood of the Boulevard Ring avenue

№	Name of a specie	λ_i^1				E_i^1	Rating
1	European white birch	0,285	0,399	0,138	0,113	0,349	17
2	European hawthorn	0,412	0,522	0,147	0,161	0,494	13
3	European white elm	0,508	0,690	0,224	0,133	0,602	3
4	Witch elm	0,488	0,635	0,183	0,125	0,572	7
5	Single-seed hawthorn	0,283	0,516	0,706	0,845	0,613	2
6	Cotoneaster	0,339	0,563	0,747	0,857	0,654	1
7	Norway maple	0,324	0,487	0,617	0,730	0,564	9
8	Tatarian maple	0,273	0,443	0,576	0,706	0,522	11
9	Canadian maple	0,206	0,376	0,515	0,676	0,464	16
10	Large-leaved linden	0,327	0,506	0,645	0,777	0,590	5
11	Little-leaved linden	0,230	0,399	0,532	0,686	0,483	14
12	Hungarian lilac	0,228	0,393	0,523	0,680	0,477	15
13	Common lilac	0,285	0,480	0,633	0,779	0,569	8
14	Cottonwood	0,325	0,502	0,643	0,742	0,577	6
15	Rough-bark poplar	0,300	0,469	0,603	0,716	0,546	10
16	European ash	0,318	0,507	0,653	0,788	0,592	4
17	Black ash	0,230	0,411	0,557	0,715	0,500	12

Table 5.22 Fuzzy rating points, rating points and rating of species of woody plants and brushwood of other parkways

№	Name of a specie	λ_i^2				E_i^1	Rating
1	European white birch	0,288	0,488	0,650	0,783	0,562	6
2	European hawthorn	0,200	0,353	0,486	0,643	0,429	16
3	European white elm	0,222	0,345	0,470	0,657	0,434	11
4	Witch elm	0,276	0,426	0,555	0,718	0,504	12
5	Single-seed hawthorn	0,320	0,440	0,546	0,694	0,512	10
6	Cotoneaster	0,221	0,425	0,600	0,766	0,512	9
7	Norway maple	0,408	0,581	0,710	0,817	0,639	1
8	Tatarian maple	0,256	0,486	0,680	0,838	0,574	4
9	Canadian maple	0,274	0,437	0,576	0,729	0,514	8
10	Large-leaved linden	0,381	0,537	0,657	0,785	0,602	2
11	Little-leaved linden	0,266	0,423	0,557	0,713	0,500	13
12	Hungarian lilac	0,332	0,531	0,687	0,818	0,602	3
13	Common lilac	0,317	0,483	0,614	0,754	0,552	7
14	Cottonwood	0,270	0,395	0,518	0,694	0,482	14
15	Rough-bark poplar	0,191	0,334	0,466	0,623	0,412	17
16	European ash	0,350	0,504	0,622	0,746	0,566	5
17	Black ash	0,207	0,353	0,487	0,654	0,434	15

Example 5.5. The analysis of enterprise bankruptcy risk. Let us consider characteristic D — «degree of bankruptcy risk of an enterprise» with linguistic values D_1 — "insignificant", D_2 — "low", D_3 — "mean", D_4 — "high". D_5 — "highest". This characteristic can be put in correspondence with the characteristic C — «an enterprise condition», and linguistic values D_l, $l = \overline{1,5}$ of characteristics D can be put in correspondence with linguistic values C_5 — "extreme well-being", C_4 — "rather well-being", C_3 — "mean condition", C_2 — "trouble", C_1 — "extreme trouble" of characteristic C. Variation range of the characteristic is a segment [0.1].

According to [187], membership functions of linguistic values C_l, $l = \overline{1,5}$ can be functions η_l, $l = \overline{1,5}$ according to parameters:

$$\eta_1 \equiv (0; 0,15; 0; 0,10);$$
$$\eta_2 \equiv (0,25; 0,35; 0,10; 0,10);$$
$$\eta_3 \equiv (0,45; 0,55; 0,10; 0,10);$$
$$\eta_4 \equiv (0,65; 0,75; 0,10; 0,10);$$
$$\eta_5 \equiv (0,85; 1; 0,10; 0).$$

Experts [187] established that the group of six separate indexes having equal importance for the analysis purpose is sufficient to carry out the complex analysis of the chosen enterprises with very high degree of reliability.

Such indexes are:

X_1 — equity-assets ratio (ratio of owned funds to total balance);

X_2 — net current assets/turnover assets ratio;

X_3 — quick asset ratio (sum of monetary resources and debts to short-term liabilities ratio);

X_4 — absolute liquidity ratio (monetary resources to short-term liabilities ratio);

X_5 — annual assets turnover (revenues to average annual assets ratio);

X_6 — return on assets (net profit to average annual assets ratio).

Linguistic values of characteristics X_j, $j = \overline{1,6}$ are the following values: X_{1j} — «very low», X_{2j} — "low", X_{3j} — "mean", X_{4j} — "high", X_{5j} — "very high", $j = \overline{1,6}$, with corresponding membership functions μ_{lj}, $l = \overline{1,5}$, $j = \overline{1,6}$

$$\mu_{11} \equiv (0; 0,100; 0; 0,100), \quad \mu_{21} \equiv (0,200; 0,250; 0,100; 0,050),$$
$$\mu_{31} \equiv (0,300; 0,450; 0,050; 0,050), \quad \mu_{41} \equiv (0,500; 0,600; 0,050; 0,100),$$
$$\mu_{51} \equiv (0,700; 1; 0,100; 0); \quad \mu_{12} \equiv (-1; -0,005; 0; 0,005),$$
$$\mu_{22} \equiv (0; 0,090; 0,005; 0,020), \quad \mu_{32} \equiv (0,110; 0,300; 0,020; 0,050),$$

$\mu_{42} \equiv (0,350; 0,450; 0,050; 0,050),\ \mu_{52} \equiv (0,500; 1; 0,050; 0);$

$\mu_{13} \equiv (0; 0,500; 0; 0,100),\ \mu_{23} \equiv (0,600; 0,700; 0,100; 0,100),$

$\mu_{33} \equiv (0,800; 0,900; 0,100; 0,100),\ \mu_{43} \equiv (1; 1,300; 0,100; 0,200),$

$\mu_{53} \equiv (1,500; \infty; 0,200; \infty);\ \mu_{14} \equiv (0; 0,020; 0; 0,010),$

$\mu_{24} \equiv (0,030; 0,080; 0,020; 0,020),\ \mu_{34} \equiv (0,100; 0,300; 0,020; 0,050),$

$\mu_{44} \equiv (0,350; 0,500; 0,050; 0,100),\ \mu_{54} \equiv (0,600; \infty; 0,100; \infty);$

$\mu_{15} \equiv (0; 0,120; 0; 0,020),\ \mu_{25} \equiv (0,140; 0,180; 0,020; 0,020),$

$\mu_{35} \equiv (0,200; 0,300; 0,020; 0,100),\ \mu_{45} \equiv (0,400; 0,500; 0,100; 0,300),$

$\mu_{55} \equiv (0,800; \infty; 0,300; \infty);\ \mu_{16} \equiv (-\infty; 0; -\infty; 0),$

$\mu_{26} \equiv (0; 0,006; 0; 0,004),\ \mu_{36} \equiv (0,010; 0,060; 0,004; 0,040),$

$\mu_{46} \equiv (0,100; 0,225; 0,040; 0,175),\ \mu_{56} \equiv (0,400; \infty; 0,175; \infty).$

Let us consider a machine-building enterprise to be analyzed over two periods — the third and fourth quarters of 1998 (period 1 and period 2), and characterized by values of indexes X_j, $j = \overline{1,6}$, presented in Table 5.23.

In Table 5.24 and Table 5.25 data $\mu_{lj}\left(c_j^1\right)$, $\mu_{lj}\left(c_j^2\right)$, $l = \overline{1,5}$, $j = \overline{1,6}$ are shown, accordingly.

Table 5.23 Values of financial indexes

Code number of index X_j	Name of an index value X_j	Value of index X_j — c_j^1 over the period 1	Value of index $X_j - c_j^2$ over the period 2
X_1	Equity-assets ratio	0,839	0,822
X_2	Net current assets/own means ratio	0,001	-0,060
X_3	Quick asset ratio	0,348	0,208
X_4	Absolute liquidity ratio	0,001	0,0001
X_5	Annual assets turnover	0,162	0,221
X_6	Return on assets	-4%	-4,3%

Ratings of Objects

Table 5.24 Values of $\mu_{ij}\left(c_j^1\right)$ over the period 1

Index code number	$\mu_{1j}\left(c_j^1\right)$	$\mu_{2j}\left(c_j^1\right)$	$\mu_{3j}\left(c_j^1\right)$	$\mu_{4j}\left(c_j^1\right)$	$\mu_{5j}\left(c_j^1\right)$
X_1	0	0	0	0	1
X_2	0	1	0	0	0
X_3	1	0	0	0	0
X_4	1	0	0	0	0
X_5	0	1	0	0	0
X_6	1	0	0	0	0
X_7	0,500	0,333	0	0	0,167

Table 5.25 Values of $\mu_{ij}\left(c_j^2\right)$ over the period 2

Index code number	$\mu_{1j}\left(c_j^2\right)$	$\mu_{2j}\left(c_j^2\right)$	$\mu_{3j}\left(c_j^2\right)$	$\mu_{4j}\left(c_j^2\right)$	$\mu_{5j}\left(c_j^2\right)$
X_1	0	0	0	0	1
X_2	1	0	0	0	0
X_3	1	0	0	0	0
X_4	1	0	0	0	0
X_5	0	0	1	0	0
X_6	1	0	0	0	0
X_7	0,666	0	0,167	0	0,167

According to data from Table 5.24 and Table 5.25, we obtain the value for "enterprise condition" over the period 1 is equal to 0.291, and over the period 2 equal to 0.287. Over both periods the enterprise is recognized as unsuccessful, and degree of bankruptcy risk is estimated as high, and there was some deterioration of a condition of the enterprise resulted from overlapping of qualitative growth of turnover with a qualitative falling of equity-assets ratio.

Chapter 6
Complete Orthogonal Semantic Spaces in Problems of the Fuzzy Regression Analysis

6.1 The Analysis of Known Methods of Fuzzy Regression Analysis

To analyze relations between qualitative characteristics and the prediction of their values the methods of fuzzy regression analysis are used, which being actively developed have already considerably expanded boundaries of application of classical regression analysis methods, i.e. they allow to construct the regression relations on the basis of fuzzy initial information. Besides, this information can be of both quantitative and qualitative nature, thus, making possible application of methods of fuzzy regression analysis in the theory of expert evaluations and ensuring practical applications in various spheres of human activity [188, 189].

Methods of fuzzy regression analysis are used to study behavior of complex engineering, ecological and other systems with output indexes depending on a great many of parameters [188]. These methods are applied to construct regression models not only within the limits of the fuzzy initial information, but also within the limits of the definite information. In this case the predicted output values are provided as fuzzy numbers. Such representation is explained by the fact that the real system is always more complex than any of its model not capable of combining all entering indexes on which the output index depends.

The first fuzzy linear regression model [190] excited interest in contributors, thus resulting in occurrence of new fuzzy regression models based on outcomes obtained in [191—193], and various optimizing criteria. Today, a number of linear fuzzy regression models [194-210] is developed, and approaches to building of nonlinear fuzzy regression models [210-212] are outlined. In [195, 196, 197, 199, 203, 209] optimizing criteria are constructed aimed at minimization of fuzziness of output model fuzzy values and the subsequent application of linear programming methods. In [206], based on [213, 214], interval regression model is under construction using methods considered in [203, 208, 209].

There appears to be three different approaches under the heading of "Fuzzy Regression":

(a) Methods that were proposed by H. Tanaka [190] and further elaborated in current literature [188, 189, 194, 196-212], where the coefficients of input

O.M. Poleshchuk and E.G. Komarov: Expert Fuzzy Info. Processing, STUDFUZZ 268, pp. 183–225.
springerlink.com   © Springer-Verlag Berlin Heidelberg 2011

variables are assumed to be fuzzy numbers. These fuzzy regression models are based on the possibility theory instead of the probability theory or they are based on both possibility and probability theories.

(b) Method proposed by R.J. Hathaway and J.C. Bezdek [215] where first the fuzzy clusters determined by an fuzzy c-means clustering (FCM) algorithm define how many ordinary regressions are to be constructed, one for each cluster. Next each fuzzy cluster is used essentially for switching purposes to determine the most appropriate ordinary regression that is to be applied for a new input from amongst a number of ordinary regressions determined in the first place.

(c) Methods proposed by I.B.Turksen [216] and A. Celikyilmaz [217], where the fuzzy functions approach to system modeling was developed. The new fuzzy functions approach augments the membership values together with their transformations to form a new input variable to find local functions. First the given system domain is fuzzy partitioned into c clusters using fuzzy c-means clustering algorithm. Then, one regression function is calculated to model the behavior of each partition. In [216] linear regression function to estimate the parameters of each function is proposed. A new fuzzy system modeling approach that identifies the fuzzy functions using support vector machines is proposed in [218]. This new approach is structurally different from the fuzzy rule base approaches and fuzzy regression methods. Method support vector machines is applied to determine the support vectors for each fuzzy cluster obtained by fuzzy c-means clustering algorithm. Original input variables, the membership values obtained from the fuzzy c-means clustering algorithm together with their transformations form a new augmented set of input variables. Methods proposed in [216-217], were investigated in [219].

In this chapter we have developed a linear and nonlinear regression models, belonging to group (a). The methods of fuzzy regression from this group have received a lot of developing in the past years. A major difference between fuzzy regression and ordinary regression [220] is in dealing with errors as fuzzy variables in fuzzy regression modeling, and in dealing with errors as random variables in ordinary regression modeling. The researchers have tried to integrate both fuzziness and randomness into regression model. As a result of this the hybrid fuzzy least-squares regressions were developed [189, 200—202, 204, 205, 207, 210-212].

In [200, 201] the least squares method is applied to deviations from unity of possibilities of equality of observable output normal triangular numbers and model output normal triangular numbers. As known, membership function of a normal triangular number looks like

$$\mu_{\tilde{A}}(x) = \begin{cases} 1 - \dfrac{a_1 - x}{a_L}, & 0 < \dfrac{a_1 - x}{a_L} \le 1, \ a_L > 0; \\ 1 - \dfrac{x - a_1}{a_R}, & 0 < \dfrac{x - a_1}{a_R} \le 1, \ a_R > 0; \\ 1, & x = a_1; \\ 0, & x < a_1 - a_L \ \text{or} \ x > a_1 + a_R. \end{cases}$$

$$(6.1)$$

The possibility of equality of two fuzzy numbers $\tilde{A}$ and $\tilde{B}$ with membership functions $\mu(x)$, $\eta(x)$, accordingly, is determined by the formula according to [192]:

$$\text{Pos}\left(\tilde{A} = \tilde{B}\right) = \max_x \min\left[\eta(x), \mu(x)\right].$$

In [197, 207] the least squares method is applied to deviations of centers of model output normal triangular numbers from centers of observable output normal triangular numbers. The centre of a normal triangular number with membership function (6.1) is the number a_1. In [207] the optimization problem of a relative minimum of the sum of fuzziness coefficients of output model normal triangular numbers is solved. Fuzziness coefficients of a normal triangular number are numbers a_L, a_R.

However, the methods of hybrid regression analysis as a rule are limited by consideration of linear regression models and of a slender group of membership functions (as a rule triangular fuzzy numbers are considered). Moreover, the hybrid regression analysis must provide a way to model the observed fuzzy data, such as linguistic descriptions of the type: "good", "very good", "excellent", which may be T - fuzzy numbers.

In practical problems fuzzy data with tolerance membership functions are often considered, therefore the problem of their analysis by regression analysis methods is acute enough. In connection with reviewing of the limited spectrum of membership functions of input data, a gap in methods of fuzzy regression analysis occurred which has been partially filled in [194].

The model-building method of hybrid fuzzy least-squares regression in the form of a system of classical regression equations (for each parameter of membership functions of initial fuzzy data) described here in, unlike other methods, can be applied both to unimodal, and to tolerance membership functions of input data. The method [194] limitations allow constructing regression model only with definite coefficients. Obviously, it dramatically limits possibilities of model and makes the problem of developing regression models with fuzzy coefficients a model of the day.

Methods of the fuzzy information formalization based on COSS described in Chapter 2 allow representing results of an expert evaluation of qualitative characteristics in the form of a group of fuzzy numbers explicitly considered in §2.1. Thus, these fuzzy numbers can be used as input and output data in a fuzzy regression model describing relations between estimated qualitative characteristics.

In order to include T - fuzzy numbers into a hybrid regression, a need for developing a new method exists. Therefore, a new linear and nonlinear multiple hybrid regressions are proposed and developed in this chapter. The developed methods allow to construct relations among qualitative characteristics and to predict their meanings.

6.2　A Method of Defuzzification of Fuzzy Numbers on the Basis of the Weighed Sets

Known methods of defuzzification of fuzzy numbers allow obtaining only pointwise aggregating indexes for these numbers, thereby, making indiscernible, for example, unimodal numbers with different coefficients of fuzziness or unimodal numbers with tolerance numbers. Thus, these methods essentially lose informational features of fuzzy numbers.

The development of defuzzification method is aimed in keeping, whenever possible, informational features of fuzzy numbers, and to use them for building of regression models.

In [84] for a normal triangular number $\tilde{B} \equiv (b, b_L, b_R)$ definition of the weighed point is provided

$$B = \int_0^1 [b - (1-\alpha)b_L + b + (1-\alpha)b_R]\alpha d\alpha = b + \frac{1}{6}(b_R - b_L). \tag{6.2}$$

Let us consider symmetric $(b_L = b_R)$ triangular numbers with a typical point b. Computing by the formula (6.2) weighed points for such fuzzy numbers, we obtain that irrespective of values of fuzziness coefficients b_L, b_R, the weighed points are equal to the same number b.

To have a possibility to determine various indexes for similar numbers, the authors have introduced the new concept — the weighed set.

As known,

$$A_\alpha = \{x \in R : \mu_A(x) \ge \alpha\} = [A_\alpha^1, A_\alpha^2] =$$
$$= [a_1 - L^{-1}(\alpha)a_L, a_2 + R^{-1}(\alpha)a_R], \ \alpha \in [0,1] \tag{6.3}$$

is referred to as set of α-level of Λ-tolerance number $\tilde{A} \equiv (a_1, a_2, a_L, a_R)$.

Let us extend definition of the weighed point (6.2) of normal triangular number onto Λ-unimodal numbers [93, 221].

The number

$$B = \int_0^1 [2b - L^{-1}(\alpha)b_L + R^{-1}(\alpha)b_R]\alpha d\alpha = b + rb_R - lb_L;$$

$$l = \int_0^1 L^{-1}(\alpha)\alpha d\alpha, \ r = \int_0^1 R^{-1}(\alpha)\alpha d\alpha. \tag{6.4}$$

is referred to as weighed point of Λ-unimodal number $\tilde{B} \equiv (b, b_L, b_R)$.

As described above, when operating with Λ-tolerance numbers (sometimes also with Λ-unimodal numbers) pointwise representation of fuzzy numbers is not always informative enough. For reasons given, definition of the weighed set for Λ-tolerance numbers is introduced.

A collection of the weighed points of all Λ-unimodal numbers belonging to Λ-tolerance number is referred to as the weighed set of this Λ-tolerance number. This definition is extended also on Λ-unimodal numbers.

The Proposition 6.1. Weighed set of Λ-tolerance number $\tilde{A} \equiv (a_1, a_2, a_L, a_R)$ is a segment $[A_1, A_2]$, where $A_1 = a_1 - la_L$, $A_2 = a_2 + la_R$. Let us refer this segment as the weighed segment of Λ-tolerance number $\overline{A}$.

The proof. Let us consider two unimodal fuzzy numbers $\tilde{B}_1 \equiv (a_1, a_L, 0)$, $\tilde{B}_2 \equiv (a_2, 0, a_R)$ which belong to Λ-tolerance number $\tilde{A} \equiv (a_1, a_2, a_L, a_R)$. Let us denote sets of α-level of numbers $\tilde{B}_1, \tilde{B}_2$ with $B_{1\alpha} = \left[B_{1\alpha}^1, a_1\right]$ and $B_{2\alpha} = \left[a_2, B_{2\alpha}^2\right]$, accordingly, and assign weighed points for these numbers in according with (6.4)

$$A_1 = \int_0^1 \left(B_{1\alpha}^1 + a_1\right)\alpha d\alpha = \int_0^1 \left(2a_1 - L^{-1}(\alpha)a_L\right)\alpha d\alpha = a_1 - \int_0^1 L^{-1}(\alpha)a_L\alpha d\alpha = a_1 - la_L;$$

$$A_2 = \int_0^1 \left(a_2 + B_{2\alpha}^2\right)\alpha d\alpha = \int_0^1 \left[2a_2 + R^{-1}(\alpha)a_R\right]\alpha d\alpha = a_2 + \int_0^1 R^{-1}(\alpha)a_R\alpha d\alpha = a_2 + ra_R,$$

Where $\int_0^1 L^{-1}(\alpha)\alpha d\alpha = l$, $\int_0^1 R^{-1}(\alpha)\alpha d\alpha = r$.

Let us consider arbitrary Λ-unimodal number $\tilde{B} \equiv (b, b_L, b_R)$ which belongs to tolerance number $\tilde{A} \equiv (a_1, a_2, a_L, a_R)$. Let us denote α-level set with $[B_\alpha^1, B_\alpha^2]$, and the weighed point $\tilde{B}$ with B. From definition of a membership of one fuzzy number to another it follows that

$$B_{1\alpha}^1 \leq B_\alpha^1, \ a_1 \leq B_\alpha^2, \ a_2 \geq B_\alpha^1, \ B_{2\alpha}^2 \geq B_\alpha^2$$

$$\frac{B_{1\alpha}^1 + a_1}{2} \leq \frac{B_\alpha^1 + B_\alpha^2}{2}, \ \frac{a_2 + B_{2\alpha}^2}{2} \geq \frac{B_\alpha^1 + B_\alpha^2}{2} \Rightarrow A_1 \leq B, \ A_2 \geq B.$$

The proposition 6.1 is proved.

The Proposition 6.2. Weighed segment of sum of Λ-tolerance numbers is equal to the sum of the weighed segments of these numbers.

The proof. Let us prove that the sum of Λ-tolerance numbers $\tilde{A} \equiv (a_1, a_2, a_{L_1}, a_{R_1})$, $\tilde{B} \equiv (b_1, b_2, b_{L_2}, b_{R_2})$ with the weighed segments $[A_1, A_2]$, $[B_1, B_2]$, accordingly, has the weighed segment $[A_1 + B_1, A_2 + B_2]$. Let us denote the weighed segment $\tilde{A} + \tilde{B}$ with $[C_1, C_2]$. Then in accordance with (6.4)

$$C_1 = \int_0^1 \left[2(a_1 + b_1) - L_1^{-1}(\alpha)a_{L_1} - L_2^{-1}(\alpha)b_{L_2}\right]\alpha d\alpha =$$

$$= a_1 + b_1 - l_1 a_{L_1} - l_2 b_{L_2} = A_1 + B_1;$$

$$C_2 = \int_0^1 \left[2(a_2 + b_2) + R_1^{-1}(\alpha) a_{R_1} + R_2^{-1}(\alpha) b_{R_2} \right] \alpha d\alpha =$$

$$= a_2 + b_2 + r_1 a_{R_1} + r_2 b_{R_2} = A_2 + B_2;$$

$$l_1 = \int_0^1 L_1^{-1}(\alpha) \alpha d\alpha, \quad r_1 = \int_0^1 R_1^{-1}(\alpha) \alpha d\alpha;$$

$$l_2 = \int_0^1 L_2^{-1}(\alpha) \alpha d\alpha, \quad r_2 = \int_0^1 R_2^{-1}(\alpha) \alpha d\alpha.$$

Thus $[C_1, C_2] = [A_1 + B_1, A_2 + B_2]$. The proposition 6.2 is proved.

The Proposition 6.3. Boundaries of the weighed segment of multiplication of Λ-tolerance numbers are determined by linear combinations of products of parameters of these numbers.

The proof. Let us consider fuzzy number which is the product of $\tilde{A} \equiv (a_1, a_2, a_{L_1}, a_{R_1})$ and $\tilde{B} \equiv (b_1, b_2, b_{L_2}, b_{R_2})$, and let us denote it with $\tilde{D} = \tilde{A} \times \tilde{B}$. Let us write out α-level sets $\tilde{A}$ and $\tilde{B}$ according to (6.3)

$$A_\alpha = \left[A_\alpha^1, A_\alpha^2 \right] = \left[a_1 - L_1^{-1}(\alpha) a_{L_1}, \, a_2 + R_1^{-1}(\alpha) a_{R_1} \right];$$

$$B_\alpha = \left[B_\alpha^1, B_\alpha^2 \right] = \left[b_1 - L_2^{-1}(\alpha) b_{L_2}, \, b_2 + R_2^{-1}(\alpha) b_{R_2} \right]$$

According to multiplication operation for fuzzy numbers of $\tilde{A}$ and $\tilde{B}$, α-level set $\tilde{D}$ looks like

$$D_\alpha = \left[\min \left(A_\alpha^1 B_\alpha^1, A_\alpha^1 B_\alpha^2, A_\alpha^2 B_\alpha^1, A_\alpha^2 B_\alpha^2 \right), \, \max \left(A_\alpha^1 B_\alpha^1, A_\alpha^1 B_\alpha^2, A_\alpha^2 B_\alpha^1, A_\alpha^2 B_\alpha^2 \right) \right]$$

Without limiting a generality, let us consider that $a_1 - a_{L_1} > 0$, $b + b_{R_2} < 0$ ($\tilde{A}$ — a positive number, $\tilde{B}$ — a negative number). Proofs of other cases are carried out similarly. Let us compute the weighed segment $[D_1, D_2]$ for Λ-tolerance number $\tilde{D}$ (Λ-tolerancy of number $\tilde{D}$ is proved in the proposition 2.2):

$$D_1 = \int_0^1 \left[2a_2 b_1 - a_2 b_{L_2} L_2^{-1}(\alpha) + a_{R_1} b_1 R_1^{-1}(\alpha) - a_{R_1} b_{L_2} L_2^{-1}(\alpha) R_1^{-1}(\alpha) \right] \alpha d\alpha =$$

$$= a_2 b_1 - l_2 a_2 b_{L_2} + r_1 a_{R_1} b_1 - m a_{R_1} b_{L_2};$$

$$D_2 = \int_0^1 \left[2a_1 b_2 + a_1 b_{R_2} R_2^{-1}(\alpha) - a_{L_1} b_2 L_1^{-1}(\alpha) - a_{L_1} b_{R_2} R_2^{-1}(\alpha) L_1^{-1}(\alpha) \right] \alpha d\alpha =$$

$$= a_1 b_2 + r_2 a_1 b_{R_2} - l_1 a_{L_1} b_2 - p a_{L_1} b_{R_2};$$

$$l_1 = \int_0^1 L_1^{-1}(\alpha) \alpha d\alpha, \quad r_1 = \int_0^1 R_1^{-1}(\alpha) \alpha d\alpha; \quad l_2 = \int_0^1 L_2^{-1}(\alpha) \alpha d\alpha, \quad r_2 = \int_0^1 R_2^{-1}(\alpha) \alpha d\alpha;$$

$$m = \int\limits_0^1 L_2^{-1}(\alpha)R_1^{-1}(\alpha)\alpha d\alpha, \quad p = \int\limits_0^1 L_1^{-1}(\alpha)R_2^{-1}(\alpha)\alpha d\alpha.$$

The proposition 6.3 is proved.

Propositions 6.1 – 6.3 are true for Λ-unimodal numbers with the following replacements $a_1 = a_2$, $b_1 = b_2$ accordingly.

6.3 Linear Hybrid Fuzzy Least-Squares Regression Model

In §2.1 the set of Λ-numbers, subdivided into Λ-tolerance and Λ-unimodal numbers, is described.

Let us define an affinity measure for two Λ-tolerance numbers $\tilde{A}$, $\tilde{B}$, with the weighed segments $[A_1, A_2]$, $[B_1, B_2]$

$$f(\tilde{A}, \tilde{B}) = \sqrt{(A_1 - B_1)^2 + (A_2 - B_2)^2}. \tag{6.5}$$

Let

$$\tilde{Y} = \begin{pmatrix} \tilde{Y}_1 \\ \dots \\ \tilde{Y}_n \end{pmatrix}, \quad \tilde{Y}_i \equiv (y_1^i, y_2^i, y_L^i, y_R^i), \quad i = \overline{1, n}$$

be output Λ-tolerance numbers, and

$$\tilde{X}_j = \begin{pmatrix} \tilde{X}_j^1 \\ \dots \\ \tilde{X}_j^n \end{pmatrix}, \quad \tilde{X}_j^i \equiv (x_1^{ji}, x_2^{ji}, x_L^{ji}, x_R^{ji}), \quad j = \overline{1, m}, \quad i = \overline{1, n}$$

be input Λ-tolerance numbers, and $\tilde{a}_j \equiv (b^j, b_L^j, b_R^j)$ unknown coefficients of regression model be Λ-unimodal numbers.

Relations of input and output data will be in the form [222]

$$\tilde{Y} = \tilde{a}_0 + \tilde{a}_1 \tilde{X}_1 + \dots + \tilde{a}_m \tilde{X}_m. \tag{6.6}$$

According to the definition of operations for fuzzy numbers and to the proposition 2.2, the multiplication of $\tilde{a}_j$ and $\tilde{X}_j^i$, $j = \overline{1, m}$, $i = \overline{1, n}$ gives Λ-tolerance numbers. If, for example, $\tilde{X}_j^i \equiv (x_1^{ji}, x_2^{ji}, x_L^{ji}, x_R^{ji})$, $j = \overline{1, m}$, $i = \overline{1, n}$, and $\tilde{a}_j \equiv (b^j, b_L^j, b_R^j)$ are positive fuzzy numbers $(x_1^{ji} - x_L^{ji} > 0, b^j - b_L^j > 0)$, then the multiplication of these numbers results in Λ-tolerance numbers with parameters

$$\left(x_1^{ji}b^j, x_2^{ji}b^j, x_1^{ji}b_L^j + x_L^{ji}b^j - x_L^{ji}b_L^j, \ x_2^{ji}b_R^j + x_R^{ji}b^j + x_R^{ji}b_R^j\right), \ j=\overline{1,m}, \ i=\overline{1,n}.$$

Using the proposition 6.1, let us compute the weighed segments $\left[y_1^i - ly_L^i, \ y_2^i + ry_R^i\right]$ for observable output data $\tilde{Y}_i$. Let us denote weighed segments of products of numbers $\tilde{a}_j$ and $\tilde{X}_j^i$ with

$$\left[\theta^1_{\tilde{a}_j\tilde{X}_j^i}\left(b^j, b_L^j, b_R^j\right), \theta^2_{\tilde{a}_j\tilde{X}_j^i}\left(b^j, b_L^j, b_R^j\right)\right] \ j=\overline{1,m}, \ i=\overline{1,n} \tag{6.7}$$

According to the proposition 6.3, boundaries of segments (6.7) are stipulated by linear combinations of products of parameters of fuzzy numbers $\tilde{a}_j$ and $\tilde{X}_j^i$, $j=\overline{1,m}$, $i=\overline{1,n}$. If, for example, $\tilde{X}_j^i \equiv \left(x_1^{ji}, x_2^{ji}, x_L^{ji}, x_R^{ji}\right)$ is positive fuzzy number, and $\tilde{a}_j \equiv \left(b^j, b_L^j, b_R^j\right)$ is negative fuzzy number $\left(x_1^{ji} - x_L^{ji} > 0, b^j + b_R^j < 0\right)$, then

$$\theta^1_{\tilde{a}_j\tilde{X}_j^i}\left(b^j, b_L^j, b_R^j\right) = b^j\left(x_2^{ji} + rx_R^{ji}\right) - b_L^j\left(lx_2^{ji} + mx_R^{ji}\right);$$

$$\theta^2_{\tilde{a}_j\tilde{X}_j^i}\left(b^j, b_L^j, b_R^j\right) = b^j\left(x_1^{ji} - lx_L^{ji}\right) + b_R^j\left(rx_1^{ji} - mx_L^{ji}\right);$$

$$l = \int_0^1 L^{-1}(\alpha)\alpha d\alpha, \ r = \int_0^1 R^{-1}(\alpha)\alpha d\alpha; \ m = \int_0^1 L^{-1}(\alpha)R^{-1}(\alpha)\alpha d\alpha.$$

Boundaries of the weighed segments of products $\tilde{a}_j\tilde{X}_j^i$ for other relations between $\tilde{a}_j$ and $\tilde{X}_j^i$ will be linear functions, too, from unknown parameters b^j, b_L^j, b_R^j (in the considered area of their values), but differing from the example given above by coefficients of the parameters considered.

Using propositions 2.1, 2.2, 6.1 — 6.3, let us compute the weighed segments

$$\left[b^0 - lb_L^0 + \sum_{j=1}^m \theta^1_{\tilde{a}_j\tilde{X}_j^i}\left(b^j, b_L^j, b_R^j\right), \ b^0 + rb_R^0 + \sum_{j=1}^m \theta^2_{\tilde{a}_j\tilde{X}_j^i}\left(b^j, b_L^j, b_R^j\right)\right], \ i=\overline{1,n}$$

For model output data $\hat{Y}_1 = \tilde{a}_0 + \tilde{a}_1\tilde{X}_1^i + ... + \tilde{a}_m\tilde{X}_m^i$.

Let us consider a functional

$$F = \sum_{i=1}^n f^2\left(\hat{Y}_i, \tilde{Y}_i\right),$$

which characterizes an affinity measure between initial and model output data. It is easy to demonstrate, that

$$F = \sum_{i=1}^n \left[b^0 - lb_L^0 - y_1^i + ly_L^i + \sum_{j=1}^m \theta^1_{\tilde{a}_j\tilde{X}_j^i}\left(b^j, b_L^j, b_R^j\right)\right]^2 +$$

$$+ \sum_{i=1}^{n} \left[b^0 + rb_R^0 - y_2^i - ry_R^i + \sum_{j=1}^{m} \theta_{\tilde{a}_j \tilde{x}_j}^2 \left(b^j, b_L^j, b_R^j \right) \right]^2 .$$

The optimization problem is set as follows:

$$F\left(b^j, b_L^j, b_R^j \right) = \sum_{i=1}^{n} f^2 \left(\hat{Y}_i, \tilde{Y}_i \right) \rightarrow \min,$$

$$b_L^j \geq 0, \; b_R^j \geq 0, \; j = \overline{0, m}.$$

As $\theta_{\tilde{a}_j \tilde{x}_j}^1 \left(b^j, b_L^j, b_R^j \right)$ and $\theta_{\tilde{a}_j \tilde{x}_j}^2 \left(b^j, b_L^j, b_R^j \right)$ are piecewise linear functions in the field $b_L^j \geq 0$, $b_R^j \geq 0$, $j = \overline{0, m}$, then F is piecewise differentiable function, and solutions of an optimization problem are determined by means of known methods [152].

Let initial output data $\tilde{\tilde{Y}}_i \equiv \left(y_1^i, y_2^i, y_L^i, y_R^i \right)$, $i = \overline{1, n}$ be formalizations of $\tilde{\tilde{Y}}_k \equiv \left(y_1^k, y_2^k, y_L^k, y_R^k \right)$, $k = \overline{1, p}$ of linguistic values Y_k of characteristic Y. After obtaining of model output data $\hat{Y}_i$, $i = \overline{1, n}$ a problem of prediction the linguistic values of characteristic Y or identification of each fuzzy numbers with one of fuzzy numbers $\tilde{\tilde{Y}}_k$, $k = \overline{1, p}$ occurs.

Let us denote the weighed segments of output model data $\hat{Y}_i$, $i = \overline{1, n}$

$$\left[b^0 - lb_L^0 + \sum_{j=1}^{m} \theta_{\tilde{a}_j \tilde{x}_j}^1 \left(b^j, b_L^j, b_R^j \right), b^0 + rb_R^0 + \sum_{j=1}^{m} \theta_{\tilde{a}_j \tilde{x}_j}^2 \left(b^j, b_L^j, b_R^j \right) \right], i = \overline{1, n}$$

with $\left[A_1^i, A_2^i \right]$, $i = \overline{1, n}$, and the weighed segments $\left(y_1^k - ly_L^k, y_2^k + ry_R^k \right)$ of formalizations $Y_k \equiv \left(y_1^k, y_2^k, y_L^k, y_R^k \right)$, $k = \overline{1, p}$ of linguistic values Y_k, $k = \overline{1, p}$ of characteristic Y with $\left[B_1^k, B_2^k \right]$, $k = \overline{1, p}$.

Let $f^2 \left(\hat{Y}_i, \tilde{\tilde{Y}}_k \right) = \left(A_1^i - B_1^k \right)^2 + \left(A_2^i - B_2^k \right)^2$, $i = \overline{1, n}$, $k = \overline{1, p}$. Output model value $\hat{Y}_i$ is identified with linguistic value Y_s, if

$$f^2 \left(\hat{Y}_i, \tilde{\tilde{Y}}_s \right) = \min_k f^2 \left(\hat{Y}_i, \tilde{\tilde{Y}}_k \right)$$

(6.8).

The developed model is hybrid because it includes elements of fuzzy and classical regression models. The similar combination allows defining analogue of a standard deviation for observations, analogue of determination coefficient and analogue of an evaluation of a standard error for quality assurance of regression models with fuzzy input data.

$$S = \sqrt{\frac{1}{n-1} \sum_{i=1}^{n} f^2 \left(\tilde{y}_i, \overline{\tilde{Y}} \right)}, \; \overline{\tilde{Y}} = \frac{\sum_{i=1}^{n} \tilde{\tilde{Y}}_i}{n}.$$

is referred to as analogue of a standard deviation for output observations.

$$HR^2 = \frac{\sum_{i=1}^{n} f^2\left(\hat{Y}_i, \overline{\overline{Y}}\right)}{\sum_{i=1}^{n} f^2\left(\tilde{Y}_i, \overline{\overline{Y}}\right)}.$$

is referred to as analogue of coefficient of determination.

$$HS = \sqrt{\frac{1}{n-m-1} \sum_{i=1}^{n} f^2\left(\hat{Y}_i, \tilde{Y}_i\right)}.$$

is referred to as analogue of an evaluation of a standard error.

With definite input data and definite coefficients of regression model, the developed hybrid fuzzy least-squares regression model is classical regression model, and certain analogues of indexes are a standard deviation of output data, determination coefficient and standard error evaluation, accordingly.

6.4 Linear Hybrid Fuzzy Least-Squares Regression Model on the Basis of Nonnegative T -Numbers

The attention to nonnegative T -numbers is explained by the fact that while constructing regression relations for qualitative characteristics, fuzzy numbers which are formalizations of linguistic values of these characteristics based on COSS can be considered as the initial information. As universal COSS set is the segment [0, 1] rather often, the fuzzy numbers corresponding to its terms, are nonnegative.

As known, the tolerance fuzzy number $\tilde{A}$ with membership function

$$\mu_{\tilde{A}}(x) = \begin{cases} 1 + \dfrac{x - a_1}{a_L}, & a_1 - a_L \le x < a_1, \ a_L > 0; \\ 1 - \dfrac{x - a_2}{a_R}, & a_2 < x \le a_2 + a_R, \ a_R > 0; \\ 1, & a_1 \le x \le a_2; \\ 0, & x < a_1 - a_L \ \text{or} \ x > a_2 + a_R. \end{cases}$$

is referred to as T -number (fuzzy number of T -type).

A T -number is symbolically recorded as $\tilde{A} \equiv (a_1, a_2, a_L, a_R)$. The normal triangular number is the special case of T -number with $a_1 = a_2$ and is symbolically recorded as $\tilde{A} \equiv (a_1, a_L, a_R)$. With $a_L = a_R$ the triangular number is referred to as symmetric, and with $a_L \ne a_R$ — asymmetric. A T -number is referred to as nonnegative under the condition $a_1 - a_L \ge 0$.

According to the proposition 6.1 the weighed segment of T-number $\tilde{A} \equiv (a_1, a_2, a_L, a_R)$ is the segment:

$$[A_1, A_2], A_1 = a_1 - \frac{1}{6}a_L, A_2 = a_2 + \frac{1}{6}a_R.$$

Let us define an affinity measure for two T-numbers $\tilde{A}, \tilde{B}$ with the weighed segments $[A_1, A_2], [B_1, B_2]$

$$f(\tilde{A}, \tilde{B}) = \sqrt{(A_1 - B_1)^2 + (A_2 - B_2)^2}.$$

Let

$$\tilde{Y} = \begin{pmatrix} \tilde{Y}_1 \\ \dots \\ \tilde{Y}_n \end{pmatrix}, \ \tilde{Y}_i \equiv (y_1^i, y_2^i, y_L^i, y_R^i), \ y_1^i - y_L^i \geq 0, \ i = \overline{1, n}$$

be output T-numbers, and

$$\tilde{X}_j = \begin{pmatrix} \tilde{X}_j^1 \\ \dots \\ \tilde{X}_j^n \end{pmatrix}, \ \tilde{X}_j^i \equiv (x_1^{ji}, x_2^{ji}, x_L^{ji}, x_R^{ji}), \ x_1^{ji} - x_L^{ji} \geq 0, \ j = \overline{1, m}, \ i = \overline{1, n}$$

be input T-numbers.

Relation between input and output data will be determined as

$$\tilde{Y} = \tilde{a}_0 + \tilde{a}_1 \tilde{X}_1 + \dots + \tilde{a}_m \tilde{X}_m,$$

where $\tilde{a}_j \equiv (b^j, b_L^j, b_R^j), \ j = \overline{0, m}$ are unknown coefficients of a regression model.

Let us define the weighed segments, using the proposition 6.1:

$$\left[y_1^i - \frac{1}{6}y_L^i, \ y_2^i + \frac{1}{6}y_R^i \right], \ i = \overline{1, n}$$

for observable output data $\tilde{Y}_i$. Let us denote the weighed segment of product of numbers $\tilde{a}_j$ and

$$\tilde{X}_j^i, \ j = \overline{1, m}, \ i = \overline{0, n} \text{ with}$$

$$\left[\theta^1_{\tilde{a}_j \tilde{X}_j^i} (b^j, b_L^j, b_R^j), \ \theta^2_{\tilde{a}_j \tilde{X}_j^i} (b^j, b_L^j, b_R^j) \right]$$

Let us consider nonnegative T-number $\tilde{X} \equiv (x_1, x_2, x_L, x_R), \ x_1 - x_L \geq 0$ and a triangular number $\tilde{a} \equiv (b, b_L, b_R)$.

The Proposition 6.4. Boundaries of the weighed segment $\left[\theta^1_{\tilde a \tilde X}, \theta^2_{\tilde a \tilde X}\right]$ of product of fuzzy numbers $\tilde a$ and $\tilde X$ look like

$$\theta^1_{\tilde a \tilde X} = b\left[x_q + (-1)^q \frac{1}{6} x_{M_q}\right] - b_L\left[\frac{1}{6} x_q + (-1)^q \frac{1}{12} x_{M_q}\right];$$

$$\theta^2_{\tilde a \tilde X} = b\left[x_r + (-1)^r \frac{1}{6} x_{M_r}\right] + b_R\left[\frac{1}{6} x_r + (-1)^r \frac{1}{12} x_{M_r}\right];$$

$$q = \begin{cases} 1, & b - b_L \geq 0; \\ 2, & b + b_R < 0; \end{cases} \qquad M_q = \begin{cases} L, & q = 1; \\ R, & q = 2; \end{cases}$$

$$r = \begin{cases} 2, & b - b_L \geq 0; \\ 1, & b + b_R < 0; \end{cases} \qquad M_r = \begin{cases} L, & r = 1; \\ R, & r = 2. \end{cases}$$

The proof. Let us write out α-level set $\tilde X$

$$X_\alpha = \left[X^1_\alpha, X^2_\alpha\right] = \left[x_1 - (1-\alpha)x_L, x_2 + (1-\alpha)x_R\right]$$

and α-level set $\tilde a$

$$a_\alpha = \left[a^1_\alpha, a^2_\alpha\right] = \left[b - (1-\alpha)b_L, b + (1-\alpha)b_R\right].$$

If $\tilde a \equiv (b, b_L, b_R)$ is nonnegative fuzzy number $(b - b_L \geq 0)$, then according to multiplication operation of fuzzy numbers, the α-level set $\tilde a \tilde X$ looks like $\left[A^1_\alpha, A^2_\alpha\right]$, where

$$A^1_\alpha = bx_1 - (1-\alpha)bx_L - (1-\alpha)b_L x_1 + (1-\alpha)^2 b_L x_L;$$

$$A^2_\alpha = bx_2 + (1-\alpha)bx_R + (1-\alpha)b_R x_2 + (1-\alpha)^2 b_R x_R.$$

Then

$$\theta^1_{\tilde a \tilde X} = \int_0^1 \left(bx_1 + A^1_\alpha\right)\alpha d\alpha = bx_1 - \frac{1}{6}bx_L - \frac{1}{6}b_L x_1 + \frac{1}{12}b_L x_L =$$

$$= b\left(x_1 - \frac{1}{6}x_L\right) - b_L\left(\frac{1}{6}x_1 - \frac{1}{12}x_L\right);$$

$$\theta^2_{\tilde a \tilde X} = \int_0^1 \left(bx_2 + A^2_\alpha\right)\alpha d\alpha = bx_2 + \frac{1}{6}bx_R + \frac{1}{6}b_R x_2 + \frac{1}{12}b_R x_R =$$

$$= b\left(x_2 + \frac{1}{6}x_R\right) + b_R\left(\frac{1}{6}x_2 + \frac{1}{12}x_R\right).$$

If $\tilde{a} \equiv (b, b_L, b_R)$ is negative fuzzy number $(b + b_R < 0)$, then according to multiplication operation for fuzzy numbers, the α-level set $\tilde{a}\tilde{X}$ looks like $\left[B_\alpha^1, B_\alpha^2\right]$, where

$$B_\alpha^1 = bx_2 + (1-\alpha)bx_R - (1-\alpha)b_L x_2 - (1-\alpha)^2 b_L x_R;$$

$$B_\alpha^2 = bx_1 - (1-\alpha)bx_L + (1-\alpha)b_R x_1 - (1-\alpha)^2 b_R x_L.$$

Then

$$\theta_{\tilde{a}\tilde{X}}^1 = \int_0^1 \left(bx_2 + B_\alpha^1\right)\alpha d\alpha = bx_2 + \frac{1}{6}bx_R - \frac{1}{6}b_L x_2 - \frac{1}{12}b_L x_R =$$

$$= b\left(x_2 + \frac{1}{6}x_R\right) - b_L\left(\frac{1}{6}x_2 + \frac{1}{12}x_R\right);$$

$$\theta_{\tilde{a}\tilde{X}}^2 = \int_0^1 \left(bx_1 + B_\alpha^2\right)\alpha d\alpha = bx_1 - \frac{1}{6}bx_L + \frac{1}{6}b_R x_1 - \frac{1}{12}b_R x_L =$$

$$= b\left(x_1 - \frac{1}{6}x_L\right) + b_R\left(\frac{1}{6}x_1 - \frac{1}{12}x_L\right)$$

or

$$\theta_{\tilde{a}\tilde{X}}^1 = b\left[x_q + (-1)^q \frac{1}{6}x_{M_q}\right] - b_L\left[\frac{1}{6}x_q + (-1)^q \frac{1}{12}x_{M_q}\right];$$

$$\theta_{\tilde{a}\tilde{X}}^2 = b\left[x_r + (-1)^r \frac{1}{6}x_{M_r}\right] + b_R\left[\frac{1}{6}x_r + (-1)^r \frac{1}{12}x_{M_r}\right];$$

$$q = \begin{cases} 1, & b - b_L \geq 0; \\ 2, & b + b_R < 0; \end{cases} \qquad M_q = \begin{cases} L, & q = 1; \\ R, & q = 2; \end{cases}$$

$$r = \begin{cases} 2, & b - b_L \geq 0; \\ 1, & b + b_R < 0; \end{cases} \qquad M_r = \begin{cases} L, & r = 1; \\ R, & r = 2. \end{cases}$$

The proposition 6.4 is proved.

According to the proposition 6.4, if $\tilde{a}_j \equiv (b^j, b_L^j, b_R^j)$, $j = \overline{1, m}$ is nonnegative fuzzy number $(b^j + b_L^j \geq 0)$, then

$$\theta_{\tilde{a}_j \tilde{X}_j}^1 (b^j, b_L^j, b_R^j) = b^j\left(x_1^{ji} - \frac{1}{6}x_L^{ji}\right) - b_L^j\left(\frac{1}{6}x_1^{ji} - \frac{1}{12}x_L^{ji}\right);$$

$$\theta_{\tilde{a}_j \tilde{X}_j}^2 (b^j, b_L^j, b_R^j) = b^j\left(x_2^{ji} + \frac{1}{6}x_R^{ji}\right) + b_R^j\left(\frac{1}{6}x_2^{ji} + \frac{1}{12}x_R^{ji}\right).$$

If $\tilde{a}_j \equiv (b^j, b_L^j, b_R^j)$, $j = \overline{1, m}$ is negative fuzzy number $(b^j + b_R^j < 0)$, then

$$\theta^1_{\tilde{a}_j \tilde{X}^i_j}\left(b^j, b^j_L, b^j_R\right) = b^j\left(x^{ji}_2 + \frac{1}{6}x^{ji}_R\right) - b^j_L\left(\frac{1}{6}x^{ji}_2 + \frac{1}{12}x^{ji}_R\right);$$

$$\theta^2_{\tilde{a}_j \tilde{X}^i_j}\left(b^j, b^j_L, b^j_R\right) = b^j\left(x^{ji}_1 - \frac{1}{6}x^{ji}_L\right) + b^j_R\left(\frac{1}{6}x^{ji}_1 - \frac{1}{12}x^{ji}_L\right).$$

Using propositions 2.1, 2.2, 6.1 - 6.4, let us determine the weighed segments
(Let us determine the weighed segments, using propositions 2.1, 2.2, 6.1 - 6.4)

$$\left[b^0 - \frac{1}{6}b^0_L + \sum_{j=1}^{m}\theta^1_{\tilde{a}_j\tilde{X}^i_j}\left(b^j, b^j_L, b^j_R\right), b^0 + \frac{1}{6}b^0_R + \sum_{j=1}^{m}\theta^2_{\tilde{a}_j\tilde{X}^i_j}\left(b^j, b^j_L, b^j_R\right)\right], i = \overline{1, n}$$

for model output data

$$\hat{Y}_i = \tilde{a}_0 + \tilde{a}_1\tilde{X}^i_1 + ... + \tilde{a}_m\tilde{X}^i_m.$$

Let us consider a functional

$$F = \sum_{i=1}^{n} f^2\left(\hat{Y}_i, \tilde{Y}_i\right),$$

which characterizes an affinity measure between initial and model output data. It is easy to demonstrate that

$$F = \sum_{i=1}^{n}\left[b^0 - \frac{1}{6}b^0_L - y^i_1 + \frac{1}{6}y^i_L + \sum_{j=1}^{m}\theta^1_{\tilde{a}_j\tilde{X}^i_j}\left(b^j, b^j_L, b^j_R\right)\right]^2 +$$

$$+ \sum_{i=1}^{n}\left[b^0 + \frac{1}{6}b^0_R - y^i_2 - \frac{1}{6}y^i_R + \sum_{j=1}^{m}\theta^2_{\tilde{a}_j\tilde{X}^i_j}\left(b^j, b^j_L, b^j_R\right)\right]^2.$$

The optimization problem is set as follows:

$$F\left(b^j, b^j_L, b^j_R\right) = \sum_{i=1}^{n} f^2\left(\hat{Y}_i, \tilde{Y}_i\right) \rightarrow \min;$$

$$b^j_L \geq 0, \ b^j_R \geq 0, \ j = \overline{0, m}.$$

As $\theta^1_{\tilde{a}_j\tilde{X}^i_j}\left(b^j, b^j_L, b^j_R\right)$ and $\theta^2_{\tilde{a}_j\tilde{X}^i_j}\left(b^j, b^j_L, b^j_R\right)$ are piecewise linear functions in the field $b^j_L \geq 0$, $b^j_R \geq 0$, $j = \overline{0, m}$, then F is piecewise differentiable function, and solutions of an optimization problem are found by means of known methods [152].

Under condition of nonnegative regression coefficients, the optimization problem is formulated as follows:

$$F = \sum_{i=1}^{n}\left\{b^0 - \frac{1}{6}b^0_L - y^i_1 + \frac{1}{6}y^i_L + \sum_{j=1}^{m}\left[b^j\left(x^{ji}_1 - \frac{1}{6}x^{ji}_L\right) - b^j_L\left(\frac{1}{6}x^{ji}_1 - \frac{1}{12}x^{ji}_L\right)\right]\right\}^2 +$$

$$+ \sum_{i=1}^{n}\left\{b^0 + \frac{1}{6}b^0_R - y^i_2 - \frac{1}{6}y^i_R + \sum_{j=1}^{m}\left[b^j\left(x^{ji}_2 + \frac{1}{6}x^{ji}_R\right) + b^j_R\left(\frac{1}{6}x^{ji}_2 + \frac{1}{12}x^{ji}_R\right)\right]\right\}^2 \rightarrow \min.$$

Unknown parameters b^j, b^j_L, b^j_R, $j = \overline{0,m}$ of membership functions of coefficients of regression models $\tilde{a}_j$, $j = \overline{0,m}$ are found from the system of normal equations:

$$\begin{cases} \dfrac{\partial F}{\partial b^j} = 0, & j = \overline{0,m}; \\[2mm] \dfrac{\partial F}{\partial b^j_L} = 0, & j = \overline{0,m}; \\[2mm] \dfrac{\partial F}{\partial b^j_R} = 0, & j = \overline{0,m}, \end{cases} \tag{6.9}$$

Being transformed (6.9), the system turns out

$$
\begin{aligned}
&2nb_0 + \sum_{i=1}^{n}\left\{-\frac{1}{6}b^0_L - y^i_1 + \frac{1}{6}y^i_L + \sum_{j=1}^{m}\left[b^j\left(x^{ji}_1 - \frac{1}{6}x^{ji}_L\right) - b^j_L\left(\frac{1}{6}x^{ji}_1 - \frac{1}{12}x^{ji}_L\right)\right]\right\} + \\
&+ \sum_{i=1}^{n}\left\{\frac{1}{6}b^0_R - y^i_2 - \frac{1}{6}y^i_R + \sum_{j=1}^{m}\left[b^j\left(x^{ji}_2 + \frac{1}{6}x^{ji}_R\right) + b^j_R\left(\frac{1}{6}x^{ji}_2 + \frac{1}{12}x^{ji}_L\right)\right]\right\} = 0; \\[2mm]
&nb_0 - \frac{n}{6}b^0_L + \left\{-y^i_1 + \frac{1}{6}y^i_L + \sum_{j=1}^{m}\left[b^j\left(x^{ji}_1 - \frac{1}{6}x^{ji}_L\right) - b^j_L\left(\frac{1}{6}x^{ji}_1 - \frac{1}{12}x^{ji}_L\right)\right]\right\} = 0; \\[2mm]
&nb_0 + \frac{n}{6}b^0_R + \left\{-y^i_2 - \frac{1}{6}y^i_R + \sum_{j=1}^{m}\left[b^j\left(x^{ji}_2 + \frac{1}{6}x^{ji}_R\right) + b^j_R\left(\frac{1}{6}x^{ji}_2 + \frac{1}{12}x^{ji}_L\right)\right]\right\} = 0; \\[2mm]
&\left(x^{ji}_1 + x^{ji}_2 - \frac{1}{6}x^{ji}_L + \frac{1}{6}x^{ji}_R\right)\sum_{i=1}^{n}\left\{b_0 - \frac{1}{6}b^0_L - y^i_1 + \frac{1}{6}y^i_L + \sum_{j=1}^{m}\left[b^j\left(x^{ji}_1 - \frac{1}{6}x^{ji}_L\right) - b^j_L\left(\frac{1}{6}x^{ji}_1 - \frac{1}{12}x^{ji}_L\right)\right]\right\} + \\
&+ \left(x^{ji}_1 + x^{ji}_2 - \frac{1}{6}x^{ji}_L + \frac{1}{6}x^{ji}_R\right)\sum_{i=1}^{n}\left\{b_0 + \frac{1}{6}b^0_R - y^i_2 - \frac{1}{6}y^i_R + \sum_{j=1}^{m}\left[b^j\left(x^{ji}_2 + \frac{1}{6}x^{ji}_R\right) + b^j_R\left(\frac{1}{6}x^{ji}_2 + \frac{1}{12}x^{ji}_R\right)\right]\right\} = 0; \\[2mm]
&\left(\frac{1}{6}x^{ji}_1 - \frac{1}{12}x^{ji}_L\right)\sum_{i=1}^{n}\left[b_0 - \frac{1}{6}b^0_L - y^i_1 + \frac{1}{6}y^i_L + \sum_{j=1}^{m}\left[b^j\left(x^{ji}_1 - \frac{1}{6}x^{ji}_L\right) - b^j_L\left(\frac{1}{6}x^{ji}_1 - \frac{1}{12}x^{ji}_L\right)\right]\right] = 0, j = \overline{1,m}; \\[2mm]
&\left(\frac{1}{6}x^{ji}_2 + \frac{1}{12}x^{ji}_R\right)\sum_{i=1}^{n}\left[b_0 + \frac{1}{6}b^0_R - y^i_2 - \frac{1}{6}y^i_R + \sum_{j=1}^{m}\left[b^j\left(x^{ji}_2 + \frac{1}{6}x^{ji}_R\right) + b^j_R\left(\frac{1}{6}x^{ji}_2 + \frac{1}{12}x^{ji}_R\right)\right]\right] = 0, j = \overline{1,m}.
\end{aligned}
$$

The obtained system is the system of simple equations relating to variables b^j, b^j_L, b^j_R and is solved with well-known methods.

6.5 Nonlinear Hybrid Fuzzy Least-Squares Regression Model Based on Nonnegative T-Numbers

Let us consider nonnegative T-numbers $\tilde{X} \equiv (x_1, x_2, x_L, x_R)$, $x_1 - x_L \geq 0$, $\tilde{Z} \equiv (z_1, z_2, z_L, z_R)$, $z_1 - z_L \geq 0$, a triangular number $\bar{a} \equiv (b, b_L, b_R)$, and let us prove propositions, related to characteristics of the weighed segments of results of operations with these fuzzy numbers.

The proposition 6.5. Boundaries of the weighed segment $\left[\theta^1_{\tilde{a}\tilde{X}^2}, \theta^2_{\tilde{a}\tilde{X}^2}\right]$ of product of fuzzy numbers $\tilde{a}$ and $\tilde{X}^2$ look like

$$\theta^1_{\tilde{a}\tilde{X}^2} = b\left(x_q^2 + \frac{(-1)^q}{3}x_q x_{M_q} + \frac{1}{12}x_{M_q}^2\right) - b_L\left(\frac{1}{6}x_q^2 + \frac{(-1)^q}{6}x_q x_{M_q} + \frac{1}{20}x_{M_q}^2\right);$$

$$\theta^2_{\tilde{a}\tilde{X}^2} = b\left(x_r^2 + \frac{(-1)^r}{3}x_r x_{M_r} + \frac{1}{12}x_{M_r}^2\right) + b_R\left(\frac{1}{6}x_r^2 + \frac{(-1)^r}{6}x_r^2 x_{M_r}^2 + \frac{1}{20}x_{M_r}^2\right),$$

where

$$q = \begin{cases} 1, & b - b_L \geq 0; \\ 2, & b + b_R < 0; \end{cases} \quad M_q = \begin{cases} L, & q = 1; \\ R, & q = 2; \end{cases}$$

$$r = \begin{cases} 2, & b - b_L \geq 0; \\ 1, & b + b_R < 0; \end{cases} \quad M_r = \begin{cases} L, & r = 1; \\ R, & r = 2. \end{cases}$$

The proof. Let us consider a fuzzy number which is multiplication of the fuzzy $\tilde{X} \equiv (x_1, x_2, x_L, x_R)$, $x_1 - x_L \geq 0$ by itself, and let us denote it with $\tilde{D} = \tilde{X} \times \tilde{X} = \tilde{X}^2$. Let us write out α-level set $\tilde{X}$

$$X_\alpha = \left[X_\alpha^1, X_\alpha^2\right] = \left[x_1 - (1-\alpha)x_L, x_2 + (1-\alpha)x_R\right].$$

According to multiplication operation for fuzzy numbers, the α-level set $\tilde{D}$ looks like

$$D_\alpha = \left[x_1^2 - 2(1-\alpha)x_1 x_L + (1-\alpha)^2 x_L^2, x_2^2 + 2(1-\alpha)x_2 x_R + (1-\alpha)^2 x_R^2\right]$$

If $\tilde{a} \equiv (b, b_L, b_R)$ is a nonnegative number, the α-level set $\tilde{a}\tilde{D}$ looks like $\left[C_\alpha^1, C_\alpha^2\right]$, where

$$C_\alpha^1 = b\left[x_1^2 - 2(1-\alpha)x_1 x_L + (1-\alpha)^2 x_L^2\right] -$$
$$- b_L\left[(1-\alpha)x_1^2 - 2(1-\alpha)^2 x_1 x_L + (1-\alpha)^3 x_L^2\right];$$
$$C_\alpha^2 = b\left[x_2^2 + 2(1-\alpha)x_2 x_R + (1-\alpha)^2 x_R^2\right] +$$
$$+ b_R\left[(1-\alpha)x_2^2 + 2(1-\alpha)^2 x_2 x_R + (1-\alpha)^3 x_R^2\right]$$

Then

$$\theta^1_{\tilde{a}\tilde{X}^2} = \int_0^1 \left(bx_1^2 + C_\alpha^1\right)\alpha d\alpha =$$

$$= b\left(x_1^2 - \frac{1}{3}x_1 x_L + \frac{1}{12}x_L^2\right) - b_L\left(\frac{1}{6}x_1^2 - \frac{1}{6}x_1 x_L + \frac{1}{20}x_L^2\right);$$

$$\theta^2_{\tilde{a}\tilde{X}^2} = \int_0^1 \left(bx_2^2 + C_\alpha^2 \right) \alpha d\alpha =$$

$$= b\left(x_2^2 + \frac{1}{3}x_2 x_R + \frac{1}{12}x_R^2 \right) + b_R\left(\frac{1}{6}x_2^2 + \frac{1}{6}x_2 x_R + \frac{1}{20}x_R^2 \right).$$

If $\tilde{a} \equiv (b, b_L, b_R)$ is a negative number, the α-level set $\tilde{a}\tilde{D}$ looks like $\left[E_\alpha^1, E_\alpha^2 \right]$, where

$$E_\alpha^1 = b\left[x_2^2 + 2(1-\alpha)x_2 x_R + (1-\alpha)^2 x_R^2 \right] -$$
$$- b_L\left[(1-\alpha)x_2^2 + 2(1-\alpha)^2 x_2 x_R + (1-\alpha)^3 x_R^2 \right];$$
$$E_\alpha^2 = b\left[x_1^2 - 2(1-\alpha)x_1 x_L + (1-\alpha)^2 x_L^2 \right] +$$
$$+ b_R\left[(1-\alpha)x_1^2 - 2(1-\alpha)^2 x_1 x_L + (1-\alpha)^3 x_L^2 \right]$$

Then

$$\theta^1_{\tilde{a}\tilde{X}^2} = \int_0^1 \left(bx_2^2 + E_\alpha^1 \right) \alpha d\alpha =$$

$$= b\left(x_2^2 + \frac{1}{3}x_2 x_R + \frac{1}{12}x_R^2 \right) - b_L\left(\frac{1}{6}x_2^2 + \frac{1}{6}x_2 x_R + \frac{1}{20}x_R^2 \right);$$

$$\theta^2_{\tilde{a}\tilde{X}^2} = \int_0^1 \left(bx_1^2 + E_\alpha^2 \right) \alpha d\alpha =$$

$$= b\left(x_1^2 - \frac{1}{3}x_1 x_L + \frac{1}{12}x_L^2 \right) + b_R\left(\frac{1}{6}x_1^2 - \frac{1}{6}x_1 x_L + \frac{1}{20}x_L^2 \right),$$

or

$$\theta^1_{\tilde{a}\tilde{X}^2} = b\left[x_q^2 + \frac{(-1)^q}{3}x_q x_{M_q} + \frac{1}{12}x_{M_q}^2 \right] - b_L\left[\frac{1}{6}x_q^2 + \frac{(-1)^q}{6}x_q x_{M_q} + \frac{1}{20}x_{M_q}^2 \right];$$

$$\theta^2_{\tilde{a}\tilde{X}^2} = b\left[x_r^2 + \frac{(-1)^r}{3}x_r x_{M_r} + \frac{1}{12}x_{M_r}^2 \right] + b_R\left[\frac{1}{6}x_r^2 + \frac{(-1)^r}{6}x_r x_{M_r} + \frac{1}{20}x_{M_r}^2 \right],$$

where

$$q = \begin{cases} 1, & b - b_L \geq 0; \\ 2, & b + b_R < 0; \end{cases} \quad M_q = \begin{cases} L, & q = 1; \\ R, & q = 2; \end{cases}$$

$$r = \begin{cases} 2, & b - b_L \geq 0; \\ 1, & b + b_R < 0; \end{cases} \quad M_r = \begin{cases} L, & r = 1; \\ R, & r = 2. \end{cases}$$

The proposition 6.5 is proved.

The Proposition 6.6. Boundaries of the weighed segment of product of fuzzy numbers $\tilde{a}$, $\tilde{X}$ and $\tilde{Z}$ look like

$$\theta^1_{\tilde{a}\tilde{X}\tilde{Z}} = b\left[x_q z_q + \frac{(-1)^q}{6} x_q z_{M_q} + \frac{(-1)^q}{6} z_q x_{M_q} + \frac{1}{12} x_{M_q} z_{M_q} \right] -$$

$$- b_L\left[\frac{1}{6} x_q z_q + \frac{(-1)^q}{12} x_q z_{M_q} + \frac{(-1)^q}{12} z_q x_{M_q} + \frac{1}{20} x_{M_q} z_{M_q} \right];$$

$$\theta^2_{\tilde{a}\tilde{X}\tilde{Z}} = b\left[x_r z_r + \frac{(-1)^r}{6} x_r z_{M_r} + \frac{(-1)^r}{6} z_r x_{M_r} + \frac{1}{12} x_{M_r} z_{M_r} \right] +$$

$$+ b_R\left[\frac{1}{6} x_r z_r + \frac{(-1)^r}{12} x_r z_{M_r} + \frac{(-1)^r}{12} z_r x_{M_r} + \frac{1}{20} x_{M_r} z_{M_r} \right],$$

$$q = \begin{cases} 1, & b-b_L \geq 0 \\ 2, & b+b_R < 0 \end{cases}; \quad M_q = \begin{cases} L, & q=1 \\ R, & q=2 \end{cases};$$

$$r = \begin{cases} 2, & b-b_L \geq 0 \\ 1, & b+b_R < 0 \end{cases}; \quad M_r = \begin{cases} L, & r=1 \\ R, & r=2 \end{cases}.$$

The proof. Let us consider a fuzzy number which is product of fuzzy number $\tilde{X} \equiv (x_1, x_2, x_L, x_R)$ by fuzzy number $\tilde{Z} \equiv (z_1, z_2, z_L, z_R)$, and let us denote it with $\tilde{G} = \tilde{X} \times \tilde{Z}$. Let us write out α-level sets $\tilde{X}$ and $\tilde{Z}$

$$X_\alpha = \left[X^1_\alpha, X^2_\alpha \right] = \left[x_1 - (1-\alpha)x_L, \ x_2 + (1-\alpha)x_R \right];$$

$$Z_\alpha = \left[Z^1_\alpha, Z^2_\alpha \right] = \left[z_1 - (1-\alpha)z_L, \ z_2 + (1-\alpha)z_R \right].$$

As $\tilde{X}$ and $\tilde{Z}$ are nonnegative fuzzy numbers, the α-level set of fuzzy number G looks like $G_\alpha = \left[G^1_\alpha, G^2_\alpha \right]$, where

$$G^1_\alpha = x_1 z_1 - (1-\alpha)x_1 z_L - (1-\alpha)z_1 x_L + (1-\alpha)^2 x_L z_L;$$

$$G^1_\alpha = x_2 z_2 + (1-\alpha)x_2 z_R + (1-\alpha)z_2 x_R + (1-\alpha)^2 x_R z_R.$$

If $\tilde{a} \equiv (b, b_L, b_R)$ is nonnegative number, the α-level set $\tilde{a}\tilde{D}$ looks like $\left[G^1_\alpha, G^2_\alpha \right]$, where

$$G^1_\alpha = b\left[x_1 z_1 - (1-\alpha)x_1 z_L - (1-\alpha)z_1 x_L + (1-\alpha)^2 x_L z_L \right] -$$

$$- b\left[(1-\alpha)x_1 z_1 - (1-\alpha)^2 x_1 z_L - (1-\alpha)^2 z_1 x_L + (1-\alpha)^3 x_L z_L \right];$$

$$G^2_\alpha = b\left[x_2 z_2 + (1-\alpha)x_2 z_R + (1-\alpha)z_2 x_R + (1-\alpha)^2 x_R z_R \right] +$$

$$+ b_R\left[(1-\alpha)x_2 z_2 + (1-\alpha)^2 x_2 z_R + (1-\alpha)^2 z_2 x_R + (1-\alpha)^3 x_R z_R \right].$$

Then

$$\theta^1_{\tilde{a}\tilde{X}\tilde{Z}} = \int_0^1 \left(bx_1 z_1 + G^1_\alpha \right)\alpha d\alpha = b\left(x_1 z_1 - \frac{1}{6}x_1 z_L - \frac{1}{6}z_1 x_L + \frac{1}{12}x_L z_L \right) -$$

$$-b_L\left(\frac{1}{6}x_1z_1-\frac{1}{12}x_1z_L-\frac{1}{12}z_1x_1+\frac{1}{20}x_Lz_L\right);$$

$$\theta^2_{\tilde{a}\tilde{X}\tilde{Z}}=\int_0^{\cdot}\left(bx_2z_2+G^2_\alpha\right)\alpha d\alpha=b\left(x_2z_2+\frac{1}{6}x_2z_R+\frac{1}{6}z_2x_R+\frac{1}{12}x_Rz_R\right)+$$

$$+b_R\left(\frac{1}{6}x_2z_2+\frac{1}{12}x_2z_R+\frac{1}{12}z_2x_R+\frac{1}{20}x_Rz_R\right).$$

If $\tilde{a}\equiv(b,b_L,b_R)$ is a negative number, the α-level set $\tilde{a}\tilde{G}$ looks like $\left[V^1_\alpha,V^2_\alpha\right]$, where

$$V^1_\alpha=b\left[x_2z_2+(1-\alpha)x_2z_R+(1-\alpha)z_2x_R+(1-\alpha)^2x_Rz_R\right]-$$
$$-b_L\left[(1-\alpha)x_2z_2+(1-\alpha)^2x_2z_R+(1-\alpha)^2z_2x_R+(1-\alpha)^3x_Rz_R\right\}$$
$$V^2_\alpha=b\left[x_1z_1-(1-\alpha)x_1z_L-(1-\alpha)z_1x_L+(1-\alpha)^2x_Lz_L\right]+$$
$$+b_R\left[(1-\alpha)x_1z_1-(1-\alpha)^2x_1z_L-(1-\alpha)^2z_1x_L+(1-\alpha)^3x_Lz_L\right]$$

Then

$$\theta^1_{\tilde{a}\tilde{X}\tilde{Z}}=\int_0^{\cdot}\left(bx_2z_2+V^1_\alpha\right)\alpha d\alpha=b\left(x_2z_2+\frac{1}{6}x_2z_R+\frac{1}{6}z_2x_R+\frac{1}{12}x_Rz_R\right)-$$

$$-b_L\left(\frac{1}{6}x_2z_2+\frac{1}{12}x_2z_R+\frac{1}{12}z_2x_R+\frac{1}{20}x_Rz_R\right);$$

$$\theta^2_{\tilde{a}\tilde{X}\tilde{Z}}=\int_0^{\cdot}\left(bx_1z_1+V^2_\alpha\right)\alpha d\alpha=b\left(x_1z_1-\frac{1}{6}x_1z_L-\frac{1}{6}z_1x_L+\frac{1}{12}x_Lz_L\right)+$$

$$+b_R\left(\frac{1}{6}x_1z_1-\frac{1}{12}x_1z_L-\frac{1}{12}z_1x_1+\frac{1}{20}x_Lz_L\right),$$

or

$$\theta^1_{\tilde{a}\tilde{X}\tilde{Z}}=b\left(x_qz_q+\frac{(-1)^q}{6}x_qz_{M_q}+\frac{(-1)^q}{6}z_qx_{M_q}+\frac{1}{12}x_{M_q}z_{M_q}\right)-$$

$$-b_L\left(\frac{1}{6}x_qz_q+\frac{(-1)^q}{12}x_qz_{M_q}+\frac{(-1)^q}{12}z_qx_{M_q}+\frac{1}{20}x_{M_q}z_{M_q}\right);$$

$$\theta^2_{\tilde{a}\tilde{X}\tilde{Z}}=b\left(x_rz_r+\frac{(-1)^r}{6}x_rz_{M_r}+\frac{(-1)^r}{6}z_rx_{M_r}+\frac{1}{12}x_{M_r}z_{M_r}\right)+$$

$$+b_R\left(\frac{1}{6}x_rz_r+\frac{(-1)^r}{12}x_rz_{M_r}+\frac{(-1)^r}{12}z_rx_{M_r}+\frac{1}{20}x_{M_r}z_{M_r}\right),$$

where

$$q=\begin{cases}1,&b-b_L\geq0;\\2,&b+b_R<0;\end{cases}\qquad M_q=\begin{cases}L,&q=1;\\R,&q=2;\end{cases}$$

$$r = \begin{cases} 2, & b - b_L \geq 0; \\ 1, & b + b_R < 0; \end{cases} \quad M_r = \begin{cases} L, & r = 1; \\ R, & r = 2. \end{cases}$$

The proposition 6.6 is proved.

It is worth mentioning that if many nonlinear classical regression models can be reduced to linear models by means of corresponding replacements, nonlinear fuzzy regression models are more complicated to reduce.

The matter is that, for example, while multiplying fuzzy numbers it is not always possible to set an analytical form for membership function of a fuzzy number which is a result out of the multiplication. Since all known linear fuzzy regression models assume such possibility, it becomes clear why it is impossible to reduce nonlinear models to linear fuzzy regression models on its own.

Let

$$\tilde{Y} = \begin{pmatrix} \tilde{Y}_1 \\ \cdots \\ \tilde{Y}_n \end{pmatrix}, \quad \tilde{Y}_i \equiv \left(y_1^i, y_2^i, y_L^i, y_R^i \right), \quad y_1^i - y_L^i \geq 0,$$

$$i = \overline{1, n} \text{ are output } T \text{-numbers},$$

$$\tilde{X}_j = \begin{pmatrix} \tilde{X}_j^1 \\ \cdots \\ \tilde{X}_j^n \end{pmatrix}, \quad \tilde{X}_j^i \equiv \left(x_1^{ji}, x_2^{ji}, x_L^{ji}, x_R^{ji} \right), \quad x_1^{ji} - x_L^{ji} \geq 0,$$

$j = \overline{1, m}$, $i = \overline{1, n}$ are input T-numbers.

Let us search relation between input and output data in the form

$$\tilde{Y} = \tilde{a}_1 \tilde{X}_1^2 + \ldots + \tilde{a}_m \tilde{X}_m^2 + \tilde{a}_{m+1} \tilde{X}_1 \tilde{X}_2 + \ldots + \tilde{a}_{\frac{m(m+1)}{2}} \tilde{X}_{m-1} \tilde{X}_m +$$

$$+ \tilde{a}_{\frac{m^2+m+2}{2}} \tilde{X}_1 + \ldots + \tilde{a}_{\frac{m(m+3)}{2}} \tilde{X}_m + \tilde{a}_0,$$

where $\tilde{a}_k \equiv \left(b^k, b_L^k, b_R^k \right)$; $k = 0, \overline{\frac{m(m+3)}{2}}$ are unknown coefficients of regression model, and also triangular numbers.

Let us determine the weighed segments

$$\left[y_1^i - \frac{1}{6} y_L^i, \ y_2^i + \frac{1}{6} y_R^i \right], \quad i = \overline{1, n}$$

for observable output data $\tilde{Y}$.

Let us denote the weighed segment of product of numbers $\tilde{a}_k$ and $\tilde{X}_j^i$ with

$$\left[\theta_{\tilde{a}_k \tilde{X}_j^i}^1 \left(b^k, b_L^k, b_R^k \right), \ \theta_{\tilde{a}_k \tilde{X}_j^i}^2 \left(b^k, b_L^k, b_R^k \right) \right]$$

where

$$k = \overline{\frac{m^2 + m + 2}{2}, \frac{m(m+3)}{2}}, \quad j = \overline{1,m}, \ i = \overline{1,n},$$

and the weighed segment of product of numbers $\tilde{a}_k$ and $\tilde{X}_j^i$, where $k = \overline{1,m}$, $j = \overline{1,m}$, $i = \overline{1,n}$, with

$$\left[\theta^1_{\tilde{a}_k \tilde{X}_j^{i2}}\left(b^k, b_L^k, b_R^k\right), \theta^2_{\tilde{a}_k \tilde{X}_j^{i2}}\left(b^k, b_L^k, b_R^k\right)\right],$$

and the weighed segment of product of numbers $\tilde{a}_k$ and $\tilde{X}_p^i, \tilde{X}_j^i$ with

$$\left[\theta^1_{\tilde{a}_k \tilde{X}_p^i \tilde{X}_j^i}\left(b^k, b_L^k, b_R^k\right), \theta^2_{\tilde{a}_k \tilde{X}_p^i \tilde{X}_j^i}\left(b^k, b_L^k, b_R^k\right)\right]$$

where

$$k = m+1, \overline{\frac{m(m+1)}{2}}, \quad p = \overline{1, m-1}, \ j = \overline{2, m}, \ p \neq j, \ p < j, \ i = \overline{2, n}.$$

According to the proposition 6.4,

$$\theta^1_{\tilde{a}_k \tilde{X}_j^i}\left(b^k, b_L^k, b_R^k\right) = b^k\left[x_q^{ji} + (-1)^q \frac{1}{6} x_{M_q}^{ji}\right] - b_L^k\left[\frac{1}{6}x_q^{ji} + (-1)^q \frac{1}{12}x_{M_q}^{ji}\right];$$

$$\theta^2_{\tilde{a}_k \tilde{X}_j^i}\left(b^k, b_L^k, b_R^k\right) = b^k\left[x_r^{ji} + (-1)^r \frac{1}{6} x_{M_r}^{ji}\right] + b_R^k\left[\frac{1}{6}x_r^{ji} + (-1)^r \frac{1}{12}x_{M_r}^{ji}\right],$$

where

$$k = \overline{\frac{m^2 + m + 2}{2}, \frac{m(m+3)}{2}}; \quad j = \overline{1,m}; \ i = \overline{2,n}; \ q = \begin{cases} 1, & b^k - b_L^k \geq 0 \\ 2, & b^k + b_R^k < 0 \end{cases},$$

$$M_q = \begin{cases} L, & q=1 \\ R, & q=2 \end{cases}; \ r = \begin{cases} 2, & b^k - b_L^k \geq 0 \\ 1, & b^k + b_R^k < 0 \end{cases}; \ M_r = \begin{cases} L, & r=1 \\ R, & r=2 \end{cases}.$$

According to the proposition 6.5

$$\theta^1_{\tilde{a}_k \tilde{X}_j^{i2}}\left(b^k, b_L^k, b_R^k\right) = b^k\left[x_q^{ji2} + \frac{(-1)^q}{3} x_q^{ji} x_{M_q}^{ji} + \frac{1}{12}x_{M_q}^{ji2}\right] -$$

$$- b_L^k\left[\frac{1}{6}x_q^{ji2} + \frac{(-1)^q}{6} x_q^{ji} x_{M_q}^{ji} + \frac{1}{6}x_{M_q}^{ji2}\right];$$

$$\theta^2_{\tilde{a}_k \tilde{X}^2_j}\left(b^k, b^k_L, b^k_R\right) = b^k\left[x_r^{ji^2} + \frac{(-1)^r}{3}x_r^{ji}x_{M_r}^{ji} + \frac{1}{12}x_{M_r}^{ji^2}\right] + $$

$$+ b^k_R\left[\frac{1}{6}x_r^{ji^2} + \frac{(-1)^r}{6}x_r^{ji}x_{M_r}^{ji} + \frac{1}{20}x_{M_r}^{ji^2}\right],$$

where

$$k = \overline{1,m}; \quad j = \overline{1,m}; \quad i = \overline{1,n};$$

$$q = \begin{cases} 1, & b^k - b^k_L \geq 0 \\ 2, & b^k + b^k_R < 0 \end{cases}; \quad M_q = \begin{cases} L, & q = 1 \\ R, & q = 2 \end{cases};$$

$$r = \begin{cases} 2, & b^k - b^k_L \geq 0 \\ 1, & b^k + b^k_R < 0 \end{cases}; \quad M_r = \begin{cases} L, & r = 1 \\ R, & r = 2 \end{cases}.$$

According to the proposition 6.6

$$\theta^1_{\tilde{a}_k \tilde{X}^i_j \tilde{X}^i_p}\left(b^k, b^k_L, b^k_R\right) = b^k\left[x_q^{ji}x_q^{pi} + \frac{(-1)^q}{6}x_q^{ji}x_{M_q}^{pi} + \frac{(-1)^q}{6}x_q^{pi}x_{M_q}^{ji} + \frac{1}{12}x_{M_q}^{ji}x_{M_q}^{pi}\right] - $$

$$- b^k_L\left[\frac{1}{6}x_q^{ji}x_q^{pi} + \frac{(-1)^q}{12}x_q^{ji}x_{M_q}^{pi} + \frac{(-1)^q}{12}x_q^{pi}x_{M_q}^{ji} + \frac{1}{20}x_{M_q}^{ji}x_{M_q}^{pi}\right];$$

$$\theta^2_{\tilde{a}_k \tilde{X}^i_j \tilde{X}^i_p}\left(b^k, b^k_L, b^k_R\right) = b^k\left[x_r^{ji}x_r^{pi} + \frac{(-1)^r}{6}x_r^{ji}x_{M_r}^{pi} + \frac{(-1)^r}{6}x_r^{pi}x_{M_r}^{ji} + \frac{1}{12}x_{M_r}^{ji}x_{M_r}^{pi}\right] + $$

$$+ b^k_R\left[\frac{1}{6}x_r^{ji}x_r^{pi} + \frac{(-1)^r}{12}x_r^{ji}x_{M_r}^{pi} + \frac{(-1)^r}{12}x_r^{pi}x_{M_r}^{ji} + \frac{1}{20}x_{M_r}^{ji}x_{M_r}^{pi}\right],$$

where

$$k = m+1, \overline{\frac{m(m+1)}{2}}; \quad p = \overline{1,m-1}, \quad j = \overline{2,m}, \quad p \neq j, \quad p < j, \quad i = \overline{2,n};$$

$$q = \begin{cases} 1, & b^k - b^k_L \geq 0 \\ 2, & b^k + b^k_R < 0 \end{cases}; \quad M_q = \begin{cases} L, & q = 1 \\ R, & q = 2 \end{cases};$$

$$r = \begin{cases} 2, & b^k - b^k_L \geq 0 \\ 1, & b^k + b^k_R < 0 \end{cases}; \quad M_r = \begin{cases} L, & r = 1 \\ R, & r = 2 \end{cases}.$$

Let us determine the weighed segments $\left[\theta^1_{\tilde{Y}_i}, \theta^2_{\tilde{Y}_i}\right]$, $i = \overline{1,n}$ for model output data

$$\widehat{Y}_i = \tilde{a}_1 \tilde{X}_1^{i^2} + \ldots + \tilde{a}_m \tilde{X}_m^{i^2} + \tilde{a}_{m+1} \tilde{X}_1^i \tilde{X}_2^i + \ldots + \tilde{a}_{\frac{m(m+1)}{2}} \tilde{X}_{m-1}^i \tilde{X}_m^i +$$

$$+ \tilde{a}_{\frac{m^2+m+2}{2}} \tilde{X}_1^i + \ldots + \tilde{a}_{\frac{m(m+3)}{2}} \tilde{X}_m^i + \tilde{a}_0;$$

$$\theta_{\widehat{Y}_i}^1 \left(b^k, b_L^k, b_R^k \right) = b^0 - \frac{1}{6} b_L^0 + \sum_{j=1}^m \theta_{\tilde{a}_j \tilde{X}_j^{i^2}}^1 \left(b^j, b_L^j, b_R^j \right) +$$

$$+ \sum_{j=2}^m \theta_{\tilde{a}_{m+j-1} \tilde{X}_1^i \tilde{X}_j^i}^1 \left(b^{m+j-1}, b_L^{m+j-1}, b_R^{m+j-1} \right) +$$

$$+ \sum_{j=3}^m \theta_{\tilde{a}_{2m-j+3} \tilde{X}_2^i \tilde{X}_j^i}^1 \left(b^{2m-j+3}, b_L^{2m-j+3}, b_R^{2m-j+3} \right) + \ldots +$$

$$\theta_{\tilde{a}_{\frac{m^2+m}{2}} \tilde{X}_{m-1}^i \tilde{X}_m^i}^1 \left(b^{\frac{m^2+m}{2}}, b_L^{\frac{m^2+m}{2}}, b_R^{\frac{m^2+m}{2}} \right) +$$

$$+ \sum_{j=1}^m \theta_{\tilde{a}_{\frac{m^2+m}{2}+j} \tilde{X}_j^i}^1 \left(b^{\frac{m^2+m}{2}+j}, b_L^{\frac{m^2+m}{2}+j}, b_R^{\frac{m^2+m}{2}+j} \right);$$

where

$$k = \overline{0, \frac{m(m+3)}{2}};$$

$$\theta_{\widehat{Y}_i}^2 \left(b^k, b_L^k, b_R^k \right) = b^0 + \frac{1}{6} b_R^0 + \sum_{j=1}^m \theta_{\tilde{a}_j \tilde{X}_j^{i^2}}^2 \left(b^j, b_L^j, b_R^j \right) +$$

$$+ \sum_{j=2}^m \theta_{\tilde{a}_{m+j-1} \tilde{X}_1^i \tilde{X}_j^i}^2 \left(b^{m+j-1}, b_L^{m+j-1}, b_R^{m+j-1} \right) +$$

$$+ \sum_{j=3}^m \theta_{\tilde{a}_{2m-j+3} \tilde{X}_2^i \tilde{X}_j^i}^2 \left(b^{2m-j+3}, b_L^{2m-j+3}, b_R^{2m-j+3} \right) + \ldots +$$

$$\theta_{\tilde{a}_{\frac{m^2+m}{2}} \tilde{X}_{m-1}^i \tilde{X}_m^i}^2 \left(b^{\frac{m^2+m}{2}}, b_L^{\frac{m^2+m}{2}}, b_R^{\frac{m^2+m}{2}} \right) +$$

$$+ \sum_{j=1}^m \theta_{\tilde{a}_{\frac{m^2+m}{2}+j} \tilde{X}_j^i}^2 \left(b^{\frac{m^2+m}{2}+j}, b_L^{\frac{m^2+m}{2}+j}, b_R^{\frac{m^2+m}{2}+j} \right).$$

where

$$k = \overline{0, \frac{m(m+3)}{2}}.$$

Let us consider a functional

$$F\left(b^k, b_L^k, b_R^k\right) = \sum_{i=1}^{n} f^2\left(\hat{Y}_i, \tilde{Y}_i\right),$$

which characterizes an affinity measure between initial and model output data:

$$F\left(b^k, b_L^k, b_R^k\right) = \sum_{i=1}^{n} \left\{ \left[\theta_{\hat{Y}_i}^1\left(b^k, b_L^k, b_R^k\right) - y_1^i + \frac{1}{6} y_L^i \right]^2 + \right.$$

$$\left. + \left[\theta_{\hat{Y}_i}^2\left(b^k, b_L^k, b_R^k\right) - y_2^i - \frac{1}{6} y_R^i \right]^2 \right\}.$$

The optimization problem is set as follows:

$$F\left(b^k, b_L^k, b_R^k\right) = \sum_{i=1}^{n} f^2\left(\hat{Y}_i, \tilde{Y}_i\right) \to \min,$$

$$b_L^k \geq 0, \ b_R^k \geq 0, \ k = 0, \frac{\overline{m(m+3)}}{2}.$$

Since $\theta_{\hat{Y}_i}^1\left(b^k, b_L^k, b_R^k\right)$ and $\theta_{\hat{Y}_i}^2\left(b^k, b_L^k, b_R^k\right)$ are piecewise linear functions in the field

$$b_L^k \geq 0, \ b_R^k \geq 0, \ k = 0, \frac{\overline{m(m+3)}}{2},$$

then F is piecewise differentiable function, and solutions of an optimization problem are determined by means of known methods [152].

Let initial output data $\tilde{Y} \equiv \left(y_1^i, y_2^i, y_L^i, y_R^i\right)$, $i = \overline{1, n}$ are formalizations $\tilde{\tilde{Y}} \equiv \left(y_1^k, y_2^k, y_L^k, y_R^k\right)$, $k = \overline{1, p}$ of linguistic values Y_k of an attribute Y. In the determining model output data $\hat{Y}_i$, $i = \overline{1, n}$ a problem of predicting linguistic values of this attribute or identification of each fuzzy numbers $\hat{Y}_i$ with one of fuzzy numbers $\tilde{\tilde{Y}}_k$ arises.

Let us denote with $\left[A_1^i, A_2^i\right]$, $i = \overline{1, n}$ the weighed segments of output model data $\hat{Y}_i$, $i = \overline{1, n}$, and with $\left[B_1^k, B_2^k\right]$, $k = \overline{1, p}$ - the weighed segments of formalizations $\tilde{\tilde{Y}}_k \equiv \left(y_1^k, y_2^k, y_L^k, y_R^k\right)$, $k = \overline{1, p}$ of linguistic values Y_k, $k = \overline{1, p}$ of attribute Y.

Then

$$f^2\left(\hat{Y}_i, \tilde{\tilde{Y}}_k\right) = \left(A_1^i - B_1^k\right)^2 + \left(A_2^i - B_2^k\right)^2; \ i = \overline{1, n}; k = \overline{1, p}.$$

Output model value $\hat{Y}_i$ is identified with linguistic value Y_s, if

$$f^2\left(\hat{Y}_i, \tilde{\tilde{Y}}_s\right) = \min_k f^2\left(\hat{Y}_i, \tilde{\tilde{Y}}_k\right), k = \overline{1, p}.$$

6.6 Building of a Reference Pattern (Image) of Objects with Qualitative Characteristics Using Fuzzy Regression Models. Obtaining of Rating Points of Objects Based on the Reference Pattern

When performing comparative analysis of objects with non-numerical characteristics essential complexity is caused by lack of a reference pattern (image); in this connection labour-consuming procedures of comparing objects with each other are carried out, as a rule. If there are a lot of objects, labour input of these procedures increases manifold.

Let us consider N objects for which experts estimate appearances of qualitative characteristics X_j, $j = \overline{1,m}$ having essential impact on some final qualitative characteristic Y.

Let X_{lj}, $l = \overline{1,m_j}$ be levels of the verbal scales applied to an evaluation of characteristics X_j, $j = \overline{1,m}$, and Y_l, $l = \overline{1,k}$ be levels of the verbal scale applied to an evaluation of characteristic Y, accordingly. Levels are arranged in ascending order of intensity of appearances of these characteristics.

Let us denote relative numbers of objects of the considered group, which are referred to level X_{lj}, $l = \overline{1,m_j}$, $j = \overline{1,m}$ while estimating the characteristic X_j, $j = \overline{1,m}$, with a_l^j, $l = \overline{1,m_j}$, $j = \overline{1,m}$;

$$\sum_{l=1}^{m_j} a_l^j = 1, \quad j = \overline{1,m}.$$

Based on these data and the method described in §2.2, let us construct COSS with names X_j, $j = \overline{1,m}$ and term-sets X_{lj}, $l = \overline{1,m}$, $j = \overline{1,m}$. Let us denote with $\mu_{lj}(x)$ membership function of fuzzy number $\tilde{X}_{lj}$ corresponding to l-th term-set of j-th COSS, $l = \overline{1,m}$, $j = \overline{1,m}$. Let us denote fuzzy numbers $\tilde{X}_{lj}$, $l = \overline{1,m}$, $j = \overline{1,m}$ or their membership functions $\mu_{lj}(x)$; $l = \overline{1,m_j}$, $j = \overline{1,m}$ with object evaluations. Let us denote with $\tilde{X}_j^n$ and $\mu_j^n(x) \equiv \left(a_{j1}^n, a_{j2}^n, a_{jL}^n, a_{jR}^n\right)$, $n = \overline{1,N}$, $j = \overline{1,m}$ an evaluation of n-th unit within the limits of attribute X_j. Fuzzy number $\tilde{X}_j^n$ with membership function $\mu_j^n(x)$ is equal to one of fuzzy numbers $\tilde{X}_{lj}$, $l = \overline{1,m_j}$, $j = \overline{1,m}$. Let us denote with a_l, $l = \overline{1,k}$ relative numbers of the objects referred to level Y_l, $l = \overline{1,k}$ while estimating the characteristic Y.

Based on these data, let us construct COSS with name Y and term-set Y_l, $l = \overline{1,k}$. Let us denote with $\mu_l(x)$ membership function of fuzzy number $\tilde{Y}_l$ corresponding to term Y_l, $l = \overline{1,k}$. Let us refer fuzzy numbers $\tilde{Y}_l$, $l = \overline{1,k}$ or their membership functions $\mu_l(x)$, $l = \overline{1,k}$ as object evaluations. Among N objects we select those having evaluations $\mu_k(y_{k1}, y_{k2}, y_{kL}, y_{kR})$ obtained from experts [the higher evaluations of intensity of characteristic Y appearance]. Without limiting a generality, let us consider that they are objects with numbers $i = \overline{1,M}$ and membership functions of values of characteristics X_j, $j = \overline{1,k}$, $\{\mu_{ij}(x) \equiv (a^i_{j1}, a^i_{j2}, a^i_{jL}, a^i_{jR})\}$, $i = \overline{1,M}$, $j = \overline{1,m}$.

The reference pattern of successful object within the limits of some final characteristic is determined as a group of fuzzy sets (or their membership functions) [223] corresponding to appearance of characteristics X_j, $j = \overline{1,m}$ i.e.

$$\{\mu_j(x) \equiv (x_{j1}, x_{j2}, x_{jL}, x_{jR})\}, \quad j = \overline{1,m}.$$

The functional model of definition of a reference pattern is shown in Fig. 6.1.

Let us denote the weighed segments of fuzzy numbers with membership functions $\{\mu^i_j(x) \equiv (a^i_{j1}, a^i_{j2}, a^i_{jL}, a^i_{jR})\}$, $i = \overline{1,M}$, $j = \overline{1,m}$ with $[A^i_{j1}, A^i_{j2}]$, $i = \overline{1,M}$, $j = \overline{1,m}$. Then we obtain

$$A^i_{j1} = a^i_{j1} - \frac{1}{6} a^i_{jL}; \quad A^i_{j2} = a^i_{j2} + \frac{1}{6} a^i_{jR}; \quad i = \overline{1,M}; \quad j = \overline{1,m}.$$

Let us denote the weighed segments of fuzzy numbers with membership functions $\{\mu_j(x) \equiv (x_{j1}, x_{j2}, x_{jL}, x_{jR})\}$, $j = \overline{1,m}$ with $[B_{j1}, B_{j2}]$, $j = \overline{1,m}$. Then we'll obtain

$$B_{j1} = x_{j1} - \frac{1}{6} x_{jL}; \quad B_{j2} = x_{j2} + \frac{1}{6} x_{jR}.$$

Let us denote the weighed segment of fuzzy number with membership function $\mu_k(x) \equiv (y_{k1}, y_{k2}, y_{kL}, y_{kR})$ with $[C_1, C_2]$, and the weighed segment of fuzzy number obtained by substituting fuzzy numbers with membership functions $\mu_j(x) \equiv (x_{j1}, x_{j2}, x_{jL}, x_{jR})$, $j = \overline{1,m}$ in fuzzy regression model (linear or nonlinear) developed in §6.4 or 6.5, with $[D_1, D_2]$.

On substituting in the linear model, we obtain

$$D_1 = b^0 - \frac{1}{6} b^0_L + \sum_{j=1}^{m} \theta^1_{\tilde{a}_j \tilde{x}_j}; \quad D_2 = b^0 + \frac{1}{6} b^0_R + \sum_{j=1}^{m} \theta^2_{\tilde{a}_j \tilde{x}_j},$$

where

$$\theta^1_{\tilde{a}_j \tilde{x}_j} = b^j \left(x_{j1} - \frac{1}{6} x_{jL} \right) - b^j_L \left(\frac{1}{6} x_{j1} - \frac{1}{12} x_{jL} \right);$$

$$\theta^2_{\tilde{a}_j \tilde{x}_j} = b^j \left(x_{j2} + \frac{1}{6} x_{jR} \right) + b^j_R \left(\frac{1}{6} x_{j2} + \frac{1}{12} x_{jR} \right), \quad j = \overline{1,m}.$$

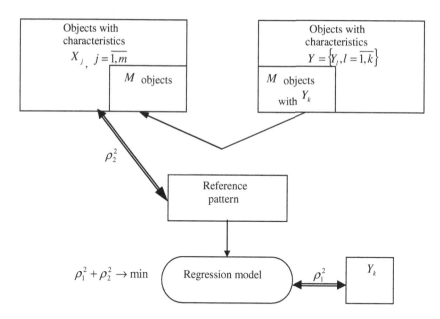

Fig. 6.1 Functional model of obtaining a reference pattern

On substituting in the linear model, we obtain

$$D_1 = b^0 - \frac{1}{6}b_L^0 + \sum_{j=1}^{m}\theta^1_{\tilde{a}_j\tilde{X}_j^2} + \sum_{j=2}^{m}\theta^1_{\tilde{a}_{m+j-1}\tilde{X}_1\tilde{X}_j} +$$

$$+ \sum_{j=3}^{m}\theta^1_{\tilde{a}_{2m-j+3}\tilde{X}_2\tilde{X}_j} + \ldots + \theta^1_{\tilde{a}_{\frac{m^2+m}{2}}\tilde{X}_{m-1}\tilde{X}_m} + \sum_{j=1}^{m}\theta^1_{\tilde{a}_{\frac{m^2+m}{2}+j}\tilde{X}_j} ;$$

$$D_2 = b^0 + \frac{1}{6}b_R^0 + \sum_{j=1}^{m}\theta^2_{\tilde{a}_j\tilde{X}_j^2} + \sum_{j=2}^{m}\theta^2_{\tilde{a}_{m+j-1}\tilde{X}_1\tilde{X}_j} +$$

$$+ \sum_{j=3}^{m}\theta^2_{\tilde{a}_{2m-j+3}\tilde{X}_2\tilde{X}_j} + \ldots + \theta^2_{\tilde{a}_{\frac{m^2+m}{2}}\tilde{X}_{m-1}\tilde{X}_m} + \sum_{j=1}^{m}\theta^2_{\tilde{a}_{\frac{m^2+m}{2}+j}\tilde{X}_j} ,$$

where

$$\theta^1_{\tilde{a}_p\tilde{X}_j} = b^p\left(x_{j1} - \frac{1}{6}x_{jL}\right) - b_L^p\left(\frac{1}{6}x_{j1} - \frac{1}{12}x_{jL}\right);$$

$$\theta^2_{\tilde{a}_p\tilde{X}_j} = b^p\left(x_{j2} + \frac{1}{6}x_{jR}\right) + b_R^p\left(\frac{1}{6}x_{j2} + \frac{1}{12}x_{jR}\right)$$

$$\text{At } p = \overline{\frac{m^2+m+2}{2}, \frac{m(m+3)}{2}}$$

and

$$\theta^1_{\tilde{a}_p \tilde{X}^2_j} = b^p \left(x^2_{j1} - \frac{1}{3} x_{j1} x_{jL} + \frac{1}{12} x^2_{jL} \right) - b^p_L \left(\frac{1}{6} x^2_{j1} - \frac{1}{6} x_{j1} x_{jL} + \frac{1}{20} x^2_{jL} \right);$$

$$\theta^2_{\tilde{a}_p \tilde{X}^2_j} = b^p \left(x^2_{j2} + \frac{1}{3} x_{j2} x_{jR} + \frac{1}{12} x^2_{jR} \right) + b^p_R \left(\frac{1}{6} x^2_{j2} + \frac{1}{6} x_{j2} x_{jR} + \frac{1}{20} x^2_{jR} \right)$$

with $p = \overline{1,m}$, $j = \overline{1,m}$, and

$$\theta^1_{\tilde{a}_p \tilde{X}_j \tilde{X}_t} = b^p \left(x_{j1} x_{t1} - \frac{1}{6} x_{j1} x_{tL} - \frac{1}{6} x_{t1} x_{jL} + \frac{1}{12} x_{jL} x_{tL} \right) -$$

$$- b^p_L \left(\frac{1}{6} x_{j1} x_{t1} - \frac{1}{12} x_{j1} x_{tL} - \frac{1}{12} x_{t1} x_{jL} + \frac{1}{20} x_{jL} x_{tL} \right);$$

$$\theta^2_{\tilde{a}_p \tilde{X}_j \tilde{X}_t} = b^p \left(x_{j2} x_{t2} + \frac{1}{6} x_{j2} x_{tR} + \frac{1}{6} x_{t2} x_{jR} + \frac{1}{12} x_{jR} x_{tR} \right) +$$

$$+ b^p_R \left(\frac{1}{6} x_{j2} x_{t2} + \frac{1}{12} x_{j2} x_{tR} + \frac{1}{12} x_{t2} x_{jR} + \frac{1}{20} x_{jR} x_{tR} \right)$$

with $p = \overline{m+1, \frac{m(m+1)}{2}}$, $t = \overline{1, m-1}$, $j = \overline{2, m}$, $t \neq j$, $t < j$.

Let us denote $(C_1 - D_1)^2 + (C_2 - D_2)^2$ with ρ^2_1, and

$$\sum_{i=1}^{M} \sum_{j=1}^{m} \left[\left(A^i_{j1} - B^i_{j1} \right)^2 + \left(A^i_{j2} - B^i_{j2} \right)^2 \right]$$

with ρ^2_1.

Unknown parameters of membership functions $\{ \mu_j(x) \equiv (x_{j1}, x_{j2}, x_{jL}, x_{jR}) \}$ of the reference pattern are found from a solution of an optimization problem

$$\rho^2_1 + \rho^2_2 \to \min$$

Under conditions $x_{j1} - x_{jL} \geq 0$, $x_{j2} + x_{jR} \leq 1$, $x_{jL} \geq 0$, $x_{jR} \geq 0$, $j = \overline{1,m}$.

We offer to use the reference pattern given in the form of a group of fuzzy numbers $\{ \mu_j(x) \equiv (x_{j1}, x_{j2}, x_{jL}, x_{jR}) \}$, $j = \overline{1,m}$ for the comparative analysis of real evaluations of objects with the pattern, obtaining of rating points and development of the operating actions aimed at success of objects within the limits of the final characteristic Y.

Let us denote the weighed segments of evaluations of n-th object [or fuzzy numbers X^n_j, $n = \overline{1,N}$ with membership functions $\{ \mu^n_j(x) \equiv (a^n_{j1}, a^n_{j2}, a^n_{jL}, a^n_{jR}) \}$, $n = \overline{1,N}$, $j = \overline{1,m}$] with $[A^n_{j1}, A^n_{j2}]$, $j = \overline{1,m}$, and

$$\sqrt{ \frac{1}{2} \left(B_{j1} - A^n_{j1} \right)^2 + \frac{1}{2} \left(B_{j2} - A^n_{j2} \right)^2 },$$

where $\lfloor B_{j1}, B_{j2} \rfloor$, $j = \overline{1,m}$ are the weighed segments of the fuzzy numbers making in aggregate a reference pattern of objects, with ρ_j, $j = \overline{1,m}$.

Let us obtain rating points of n-th object, $n = \overline{1,N}$ as follows [223, 224]:

$$r_n = 1 - \sum_{j=1}^{m} \omega_j \rho_j. \tag{6.10}$$

The rating points obtained by the formula (6.10) will differ from the evaluations obtained by a principle "the higher individual indexes are, the higher rating points are" [225—226]. The rating points constructed on the basis of s reference pattern, are quantity indexes of affinity of indexes of estimated objects and indexes of the pattern constructed on the basis of real a posteriori data. Therefore, the analysis of the objects' functioning monitoring data based on the obtained rating points improve reliability of the prediction of final indexes (in our case, they are indexes of characteristic Y) and control operations on their improving.

6.7 Examples of Fuzzy Regression Models Employment

Example 6.1. Prediction of progress of the trainees. The comparative analysis of combined (hybrid) and classical regression models. To construct classical and hybrid linear regression models, data of progress of 30 trainees in four subjects [227] are taken. From these data non-repeating results are selected and summarized in Table 6.1.

Based on the data obtained by a method described in §2.2, four COSS's are constructed, their membership function parameters are summarized in Table 6.2.

By the method described in §6.4, linear hybrid regression model is constructed with definite coefficients:

$$\tilde{Y} = a_1 \tilde{X}_1 + a_2 \tilde{X}_2 + a_3 \tilde{X}_3 + a_0,$$

where a_j, $j = \overline{0,3}$ are unknown coefficients of regression model. The solution of an optimization problem allows obtaining the model

$$\tilde{Y} = 0{,}352 \tilde{X}_1 + 0{,}466 \tilde{X}_2 + 0{,}133 \tilde{X}_3;$$

$$S = 0{,}454; \ HR = 0{,}805; \ HS = 0{,}239.$$

According to the method described in §6.4, linear hybrid regression model is constructed with fuzzy coefficients

$$\tilde{Y} = \tilde{a}_0 + \tilde{a}_1 \tilde{X}_1 + \tilde{a}_2 \tilde{X}_2 + \tilde{a}_3 \tilde{X}_3,$$

where $\tilde{a} \equiv (b^j, b_L^j, b_R^j)$, $j = \overline{0,3}$ are unknown coefficients of regression model, and also normal triangular numbers. The solution of an optimization problem allows obtaining the model

$$\tilde{Y} = (0;0,566;0) + (0,412;0,104;0)\tilde{X}_1 + (0,466;0;0)\tilde{X}_2 +$$
$$+ (0,130;0;0)\tilde{X}_3;$$
$$S = 0,454; \; HR = 0,827; \; HS = 0,213.$$

As one can see form the building process, fuzzy regression model with fuzzy coefficients has more accurate indexes then regression model with definite coefficients.

The classical linear regression model is constructed

$$Y = -0.708 + 0,301X_1 + 0,428X_2 + 0,394X_3;$$
$$S_Y = 0,949; \; R = 0,808; \; S = 0,509.$$

Table 6.1 Progress data

No.	X_1	X_2	X_3	Y
1	2	3	3	2
2	3	4	3	3
3	3	2	2	2
4	4	4	3	4
5	5	4	5	5
6	4	3	3	3
7	5	5	4	4
8	5	3	4	3
9	2	3	3	2
10	3	4	3	3

Table 6.2 Parameters of membership functions of evaluations in subjects

Term No.	X_1	X_2	X_3	Y
1	(0;0,10;0;0,10)	(0;0,05;0;0,10)	(0;0,15;0;0,20)	(0;0,10;0;0,15)
2	(0,20;0,40;0,10;0,30)	(0,15;0,45;0,10;0,30)	(0,35;0,20;0,30)	(0,25;0,60;0,15;0,10)
3	(0,70;0,80;0,30;0,15)	(0,75;0,85;0,30;0,10)	(0,65;0,85;0,30;0,05)	(0,70;0,90;0,10;0,05)
4	(0,95;1,00;0,15;0)	(0,95;1,00;0,10;0)	(0,90;1,00;0,05;0)	(0,95;1,00;0,05;0)

Considering an incorrectness of operating with elements of ordinal scales, we may assume that the constructed model does not initially have essential and substantial sense. However, in practice similar models are applied rather often. Therefore, the classical model is constructed to compare its determination coefficient with values of analogue determination coefficient for the fuzzy hybrid regressions (other quality indexes are incomparable), and also to compare the predicted output indexes obtained within the limits of all models. At this stage of investigation it is clear that fuzzy hybrid regression models have great values of determination coefficient analogue in comparison with coefficient of determination of a classical model.

Results of comparison of real output data with the output data obtained within the limits of all models are summarized in Table 6.3.

To recognize fuzzy output values indexes (6.8) are used.

As one can see from Table 6.3, the model output data obtained based on the fuzzy hybrid regression of model with fuzzy coefficients coincide with real output data in 90 %. The model output data obtained based on the fuzzy hybrid regression of model with definite coefficients, coincide with real output data in 80 %.

Table 6.3 Real and model output data

Item No.	Initial data	Model value of classical regression	Model value of fuzzy regression	
			With definite coefficients	With fuzzy coefficients
1	2	2	3	2
2	3	3	3	3
3	2	2	2	2
4	4	3	4	4
5	5	4	4	4
6	3	3	3	3
7	4	5	4	4
8	3	4	3	3
9	3	3	3	3
10	4	4	4	4

Also, the model output data obtained based on the classical regression of model coincide with real output data in 60 %.

Thus, the carried out analysis of quality classical linear and fuzzy linear hybrid regression models (constructed within the limits of the information of educational process) allows drawing a conclusion about advantages of the developed fuzzy hybrid regression model, and reasonably recommend application of this model for processing information of educational process and its reliable prediction.

Example 6.2. Prediction of commercial success of software products. Development of managerial instructions to assure the success. As input characteristics of the software products developed for automation of retail business, bank activity, insurance business and I/C account three characteristics are taken: X_1 — modifiability, X_2 — studiability and X_3 — functionality.

Modifiability is a characteristic of software products which simplifies introducing of necessary modifications and updating and includes characteristics of expansibility, structuredness and modularity.

Studiability is a characteristic which allows minimization of efforts to study and understand software programs and documentation and includes characteristics of informativeness, clearness, structuredness and readability.

Functionality is a characteristic which shows ability of software product to fulfill a number of the functions defined in its external declaration and satisfying specified or implied demands of users.

As output characteristic success of software products Y, which includes groups of these products with customers, their salesability, and recognising by leading experts is considered.

Table 6.4 The real and formalized data of expert evaluations of software products

n	X_1	X_2	X_3	Y
		Actual data		
1	3	4	3	3
2	3	3	3	3
3	4	3	5	5
4	4	4	4	4
5	5	3	5	5
6	5	4	4	5
7	3	4	4	4
8	4	3	4	3
9	4	5	3	3
10	4	5	3	3
11	3	3	4	4
12	4	4	3	4
		Formalized data		
1	(0;0,15;0;0,3)	(0,375;0,425;0,25;0,35)	(0;0,125;0;0,25)	(0;0,075;0;0,15)
2	(0;0,15;0;0,3)	(0;0,125;0;0,25)	(0;0,125;0;0,25)	(0;0,075;0;0,15)
3	(0,45;0,55;0,3;0,3)	(0;0,125;0;0,25)	(0,85;1;0,3;0)	(0,925;1;0,25;0)
4	(0,45;0,55;0,3;0,3)	(0,375;0,425;0,25;0,35)	(0,375;0,55;0,25;0,3)	(0,225;0,675;0,15;0,25)
5	(0,85;1;0,3;0)	(0;0,125;0;0,25)	(0,85;1;0,3;0)	(0,925;1;0,25;0)
6	(0,85;1;0,3;0)	(0,375;0,425;0,25;0,35)	(0,375;0,55;0)25;0,3)	(0,925;1;0,25;0)
7	(0;0,15;0;0,3)	(0,375;0,425;0,25;0;35)	(0,375;0,55;0,25;0,3)	(0,225;0,675;0,15;0,25)

Table 6.4 (*continued*)

8	(0,45;0,55;0,3;0,3)	(0;0,125;0;0,25)	(0,375;0,55;0,25;0,3)	(0;0,075;0;0,15)
9	(0,45;0,55;0,3;0,3)	(0,775;1;0,35;0)	(0;0,125;0;0,25)	(0,225;0,675;0,15;0,25)
10	(0,45;0,55;0,3;0,3)	(0,775;1;0,35;0)	(0;0,125;0;0,25)	(0,225;0,675;0,15;0,25)
11	(0;0,15;0;0,3)	(0;0,125;0;0,25)	(0,375;0,55;0,25;0,3)	(0;0,075;0;0,15)
12	(0,45;0,55;0,3;0,3)	(0,375;0,425;0,25;0,35)	(0;0,125;0;0,25)	(0;0,075;0;0,15)

It is assumed that all characteristics have three linguistic values: "low", "mean", "high". The choice of input characteristics from a certain set offered is an individual separate problem [211] aimed at detecting those which have essential impact on the output characteristic.

Twelve selected software products have been estimated by experts, and the obtained data are formalized by the method described in §2.2. Results are summarized in Table 6.4.

By the method described in § 6.5, nonlinear hybrid fuzzy least-squares regression model is constructed:

$$\tilde{Y} = (-0,483;0;0)\tilde{X}_1^2 + (0,061;0;0)\tilde{X}_3^2 +$$
$$+ (-0,121;0;0)\tilde{X}_2\tilde{X}_3 + (1,022;1,022;0)\tilde{X}_1 +$$
$$+ (0,283;0,283;0)\tilde{X}_2 + (0,017;0;0).$$

By the method described in §6.4, linear hybrid fuzzy least-squares regression model is constructed:

$$\tilde{Y} = (0,026;0;0) + (0,067;0;0)\tilde{X}_1 +$$
$$+ (0,619;0,585;0,507)\tilde{X}_2 + (0,234;0;0)\tilde{X}_3.$$

On carrying out of the comparative analysis of real output data and the model output data obtained within the limits of two constructed models, the nonlinear regression model is selected to enable further building process.

Based on the data of Table 6.4, data X_i, $i = \overline{1,3}$ of the software products obtaining the higher marks of success (tab. 6.5), is determined.

Table 6.5 Data of the software products obtaining the higher marks of success

X_1	X_2	X_3
(0,45;0,55;0,3;0,3)	(0;0,125;0;0,25)	(0,85;1;0,3;0)
(0,85;1;0,3;0)	(0;0,125;0;0,25)	(0,85;1;0,3;0)
(0,85;1;0,3;0)	(0,375;0,425;0,25;0,35)	(0,375;0,55;0,25;0,3)

For the data of Table 6.5 the weighed segments (tab. 6.6) is obtained.

Table 6.6 The weighed segments $\left[A_{j1}^i, A_{j2}^i\right]$, $j = 1,2,3$

i	$\left[A_{11}^i, A_{12}^i\right]$	$\left[A_{21}^i, A_{22}^i\right]$	$\left[A_{31}^i, A_{32}^i\right]$
1	[0,4; 0,56]	[0,05; 0,165]	[0,8; 1]
2	[0,8; 1]	[0,05; 0,165]	[0,8; 1]
3	[0,8; 1]	[0,335; 0,483]	[0,335; 0,6]

The reference pattern (image) of successful software product is determined in the form:

$$\mu_j \equiv \left(x_{j1}, x_{j2}, x_{jL}, x_{jR}\right), \quad j = 1,2,3.$$

The weighed segments of the reference pattern are denoted with $\left[A_{j1}, A_{j2}\right]$, $j = 1,2,3$, and the weighed segment of fuzzy number with parameters (0.925; 1; 0.25; 0), formalizing the higher mark of success of software product, $C_1 = 0,883$, $C_2 = 1$, is denoted with $\left[C_1, C_2\right]$.

On substituting the reference pattern in constructed nonlinear regression model the weighed segment of the obtained result is obtained:

$$D_1 = -0,483x_{12}^2 - 0,161x_{12}x_{1R} - 0,04x_{1R}^2 + 0,161x_{31}^2 -$$
$$- 0,05x_{31}x_{3L} + 0,01x_{3L}^2 - 0,121x_{22}x_{32} - 0,02x_{22}x_{3R} -$$
$$- 0,02x_{32}x_{2R} - 0,01x_{2R}x_{3R} + 0,852x_{11} - 0,085x_{1L} +$$
$$+ 0,236x_{21} - 0,024x_{2L} + 0,017;$$
$$D_2 = -0,483x_{11}^2 + 0,161x_{11}x_{1L} - 0,04x_{1L}^2 + 0,161x_{32}^2 +$$
$$+ 0,05x_{32}x_{3R} + 0,01x_{3R}^2 - 0,121x_{21}x_{31} + 0,02x_{21}x_{3L} +$$
$$+ 0,02x_{31}x_{2L} - 0,01x_{2L}x_{3L} + 1,022x_{12} + 0,17x_{1R} +$$
$$+ 0,283x_{22} + 0,047x_{2R} + 0,017.$$

Unknown parameters of membership functions $\left\{\mu_j(x) \equiv \left(x_{j1}, x_{j2}, x_{jL}, x_{jR}\right)\right\}$ of the reference pattern are obtained from the solution of an optimization problem

$$F\left(x_{j1}, x_{j2}, x_{jL}, x_{jR}\right) = \sum_{i=1}^{3}\left[\left(A_{j1}^i - B_{j1}\right)^2 + \left(A_{j2}^i - B_{j2}\right)^2\right] +$$
$$+ \left(C_1 - D_1\right)^2 + \left(C_2 - D_2\right)^2 \rightarrow \min$$

Under conditions

$$x_{j1} - x_{jL} \geq 0; \quad x_{j2} + x_{jR} \leq 1; \quad x_{jL} \geq 0; \quad x_{jR} \geq 0; \quad j = \overline{1,3}$$

or

$$F\left(x_{j1}, x_{j2}, x_{jL}, x_{jR}\right) = \left(0,4 - x_{11} + 0,167x_{1L}\right)^2 +$$
$$+ \left(0,56 - x_{12} - 0,167x_{1R}\right)^2 + 2\left(0,8 - x_{11} + 0,167x_{1L}\right)^2 +$$

$$+ 2(1 - x_{12} - 0{,}167 x_{1R})^2 + 2(0{,}05 - x_{21} + 0{,}167 x_{2L})^2 +$$
$$+ 2(0{,}165 - x_{22} - 0{,}167 x_{2R})^2 + (0{,}335 - x_{21} + 0{,}167 x_{2L})^2 +$$
$$+ (0{,}483 - x_{22} - 0{,}167 x_{2R})^2 + 2(0{,}8 - x_{31} + 0{,}167 x_{3L})^2 +$$
$$+ 2(1 - x_{32} - 0{,}167 x_{3R})^2 + (0{,}335 - x_{31} - 0{,}167 x_{3L})^2 +$$
$$+ (0{,}6 - x_{32} - 0{,}167 x_{3R})^2 + (0{,}483 x_{12}^2 + 0{,}161 x_{12} x_{1R} +$$
$$+ 0{,}04 x_{1R}^2 - 0{,}161 x_{31}^2 + 0{,}05 x_{31} x_{3L} - 0{,}01 x_{3L}^2 +$$
$$+ 0{,}121 x_{22} x_{32} + 0{,}02 x_{22} x_{3R} + 0{,}02 x_{32} x_{3R} + 0{,}01 x_{2R} x_{3R} -$$
$$- 0{,}852 x_{11} + 0{,}085 x_{1L} - 0{,}236 x_{21} + 0{,}024 x_{2L} + 0{,}886)^2 +$$
$$+ (0{,}483 x_{11}^2 - 0{,}161 x_{11} \cdot x_{1L} + 0{,}04 x_{1L}^2 - 0{,}161 x_{32}^2 -$$
$$+ 0{,}05 x_{32} x_{3R} - 0{,}01 x_{3R}^2 + 0{,}121 x_{21} x_{31} - 0{,}02 x_{21} \cdot x_{3L} -$$
$$- 0{,}02 x_{31} \cdot x_{2L} + 0{,}01 x_{2L} \cdot x_{3L} - 1{,}022 x_{12} - 0{,}17 x_{1R} -$$
$$- 0{,}283 x_{22} - 0{,}047 x_{2R} + 0{,}886)^2 \to \min,$$

with

$$x_{j1} - x_{jL} \geq 0; \ x_{j2} + x_{jR} \leq 1; \ x_{jL} \geq 0; \ x_{jR} \geq 0; \ j = \overline{1,3}.$$

The solution of the optimization problem determines the reference pattern of successful software product:

$$\mu_1(x) \equiv (0{,}789; 0{,}789; 0{,}363; 0{,}211),$$
$$\mu_2(x) \equiv (0{,}155; 0{,}155; 0; 0{,}845),$$
$$\mu_3(x) \equiv (0{,}68; 0{,}847; 0; 0{,}153).$$

Table 6.7 The weighed segments of evaluation results for software products

n	$\left[A_{11}^n, A_{12}^n\right]$	$\left[A_{21}^n, A_{22}^n\right]$	$\left[A_{31}^n, A_{32}^n\right]$
1	[0; 0,2]	[0,333; 0,477]	[0; 0,167]
2	[0; 0,2]	[0; 0,167]	[0; 0,167]
3	[0,4; 0,6]	[0; 0,167]	[0; 8,1]
4	[0,4; 0,6]	[0,33; 0,477]	[0,33; 0,6]
5	[0; 8,1]	[0; 0,167]	[0; 8,1]
6	[0; 8,1]	[0,333; 0,477]	[0,333; 0,6]
7	[0; 0,2]	[0,333; 0,477]	[0,333; 0,6]
8	[0,4; 0,6]	[0; 0,167]	[0,333; 0,6]
9	[0,4; 0,6]	[0; 723,1]	[0; 0,167]
10	[0,4; 0,6]	[0; 0,167]	[0; 0,167]
11	[0; 0,2]	[0; 0,167]	[0,333; 0,6]
12	[0,4; 0,6]	[0,333; 0,477]	[0; 0,167]

For formalized data of Table 5.7 the weighed segments are obtained and summarized in Table 6.7.

For the reference pattern of successful software product the weighed segments are obtained:

$$[B_{11}, B_{12}] = [0,728; 0,824],$$

$$[B_{21}, B_{22}] = [0,155; 0,296],$$

$$[B_{31}, B_{32}] = [0,68; 0,873].$$

By the formula (6.10), rating points of software products and the obtained results are summarized in Table 6.8.

Upon the experts' agreement the weight coefficients ω_j, $j = \overline{1,3}$ are taken equal to 1/3.

The obtained rating points of software products were used to develop managerial instructions aimed at achieving future success of software products [228—230].

Values of rating points are grouped in three intervals [0; 0.2], [0.2; 0.6], [0; 6.1], to which linguistic values: "low", "mean", "high" are assigned, accordingly.

Table 6.8 Rating points of software products and a rating based on the reference pattern

n	Rating points	Rating
1	0,284	10
2	0,301	9
3	0,754	2
4	0,647	5
5	0,789	1
6	0,683	3
7	0,463	8
8	0,665	4
9	0,251	11—12
10	0,251	11—12
11	0,481	6
12	0,468	7

If the rating point of software product falls in the first interval [0; 0.2], the software product comes back to the beginning of previous development cycle. If the rating point of software product falls in the last interval [0.6; 1], the software product is transferred to the next development cycle. If the rating point of software product falls in the mean interval [0.2; 0.6], the aggregation operator of information will be used to develop managerial instructions.

Three variables were considered: Z_1 — distance between value of characteristic X_1 of n-th software product, and value of characteristic of the reference pattern of software products; Z_2 — distance between value of characteristic X_2 of n-th software product, and value of characteristic X_2 of the reference pattern of software products; Z_3 — distance from value of characteristic X_3 of n-th software product, $n = \overline{1,12}$ to value of characteristic X_3 of the reference pattern of software products.

Each variable accepts three linguistic values "low", "mean", "high", to which values 0, 1 and 2 are put in correspondence withy the aim to construct logic function. Numerical values of variables Z_1, Z_2, Z_3 were denoted with Z_1^n, Z_2^n, Z_3^n, $n = \overline{1,12}$, accordingly:

$$Z_1^n = \sqrt{\left(B_{11} - A_{11}^n\right)^2 + \left(B_{12} - A_{12}^n\right)^2}, \ n = \overline{1,12};$$

$$Z_2^n = \sqrt{\left(B_{21} - A_{21}^n\right)^2 + \left(B_{22} - A_{22}^n\right)^2}, \ n = \overline{1,12};$$

$$Z_3^n = \sqrt{\left(B_{31} - A_{31}^n\right)^2 + \left(B_{32} - A_{32}^n\right)^2}, \ n = \overline{1,12}.$$

The interval [0; 0.3] is put in correspondence to linguistic value "low" of variables Z_1, Z_2, Z_3; [0.3; 0.95] — to value "mean", and interval [0.95; $\sqrt{2}$] — to value "high".

On agreeing with experts, the logic function F depending on variables Z_1, Z_2, Z_3 began to take three values: "the software product is conditionally transferred to the next stage and updated routinely", "an expert is invited to update the software program", "a group of experts is invited to update the software program". Values 0, 1 and 2 were put in correspondence to these values.

The functional model of development of managerial instructions based on the rating points is given in Fig. 6.2.

Experts formulated the following entry conditions $F(Z_1 = 2) = 2$, $F(Z_2 = 2) = 2$, $F(Z_3 = 2) = 2$, and fuzzy conditions "slightly-increase" for behavior of function by each argument. These conditions were formalized by means of fuzzy relations, with the related matrixes summarized in Table 6.9.

As a result of formalization of function behavior conditions and entry conditions [230], the following matrixes of the fuzzy relations with related elements summarized in Table 6.10, were obtained.

At the cross-section of a line $i+1$ and a column $j+1$ of the matrixes (Table 6.10) there are the values of confidence degrees saying that function F will take value j with value of arguments Z_1, Z_2, Z_3, equal to i, $i = \overline{0,2}$, $j = \overline{0,2}$, accordingly.

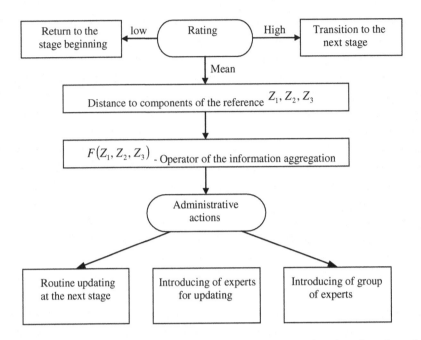

Fig. 6.2 Functional model of development of managerial instructions based on the rating points

Table 6.9 A matrix of the fuzzy relation describing "slightly-increase" of logic function F for arguments Z_1, Z_2, Z_3

п	0	1	2
F *for argument* Z_1			
0	0,9	1	0,9
1	0	0,9	1
2	0	0	0,9
F *for argument* Z_2			
0	0,7	1	0,7
1	0	0,7	1
2	0	0	0,7
F *for argument* Z_3			
0	0,8	1	0,8
1	0	0,8	1
2	0	0	0,8

Table 6.10 A matrix of the fuzzy relation describing values of logic function F for arguments Z_1, Z_2, Z_3

п	0	1	2
F for argument Z_1 (a)			
0	1	0,9	0,9
1	0,9	0,9	0,9
2	0	0	1
F for argument Z_2 (b)			
0	1	0,7	0,7
1	0,7	0,7	0,7
2	0	0	1
F for argument Z_3 (c)			
0	1	0,8	0,8
1	0,8	0,8	0,8
2	0	0	1

As a result of formalization of all conditions the relation is obtained with the matrix of 27 lines (according to number of possible values of arguments) and 3 columns (according to number of function values). Matrix elements are values of confidence degree saying that function F will accept any value depending on values of arguments Z_1, Z_2, Z_3. For example, to obtain degree of confidence saying that function F will take value 1 at $Z_1 = 0$, $Z_2 = 0$, $Z_3 = 0$, it is necessary to take a minimum of an element of the matrix 6.10 (a) placed on the cross-section of the first line and the second column, an element of a matrix 6.10 (b) placed on the cross-section of the second line and the second column, and an element of a matrix 6.10 (c) placed on the cross-section of the first line and the second column.

On completing all operations the fuzzy relation describing work of an aggregation operator of information or functions F, is obtained. Elements of a matrix of this relation are summarized in Table 6.11.

Discussing the obtained results with experts allowed obtaining the aggregation operator of information in the form of function of three-value logic with three variables. Results of work of this operator are summarized in Table 6.12.

Managerial instructions based on results of work of the operator are in the form of the following logic rules (Table 6.14).

Table 6.11 The fuzzy relation describing work of an aggregation operator of information (function F)

Argument	Function F		
	0	1	2
000	1	0,7	0,7
001	0,8	0,7	0,7
002	0	0	0,7
010	0,7	0,7	0,7
011	0,7	0,7	0,7
012	0	0	0,7
020	0	0	0,8
021	0	0	0,8
022	0	0	0,9
100	0,9	0,7	0,7
101	0,8	0,7	0,7
102	0	0	0,7
110	0,7	0,7	0,7
111	0,7	0,7	0,7
112	0	0	0,7
120	0	0	0,8
121	0	0	0,8
122	0	0	0,9
200	0	0	0,7
201	0	0	0,7
202	0	0	0,7
210	0	0	0,7
211	0	0	0,7
121	0	0	0,7
220	0	0	0,8
221	0	0	0,8
222	0	0	1

Table 6.12 Aggregation operator of information

Argument	Function
000	0
001	0
002	2
010	0
011	1

Table 6.12 (*continued*)

012	2
020	2
021	2
022	2
100	0
101	0
102	2
110	1
111	1
112	2
120	2
121	2
122	2
200	2
201	2
202	2
210	2
211	2
121	2
220	2
221	2
222	2

Table 6.13 Distances from values of software product characteristics to values of reference pattern characteristics

Product number	Z_1^n	Z_2^n	Z_3^n
1	0,914	0,254	0,98
2	0,914	0,202	0,98
7	0,914	0,254	0,442
9	0,362	0,905	0,98
10	0,362	0,905	0,98
11	0,914	0,202	0,442
12	0,362	0254	0,98

Table 6.14 Logic rules of work with software product for arguments $Z_1 - Z_3$

Rule number	Value			Administrative action
	Z_1	Z_2	Z_3	
	Low	Low	Low	
	—»—	—»—	Mean	The software product is conditionally
1	—»—	Mean	Low	transferred to the next stage and updated
	Mean	Low	—»—	routinely
	—»—	—»—	Mean	
	Low	Mean	Mean	
2	Mean	—»—	Low	An expert is invited to update software product
	—»—	—»—	Mean	
	Low	Low	High	
	—»—	Mean	—»—	
	—»—	High	Low	
	—»—	—»—	Mean	
	—»—	—»—	High	
	Mean	Low	—»—	
	—»—	Mean	—»—	
	—»—	High	Low	
	—»—	—»—	Mean	
3	—»—	—»—	High	A group of experts is invited to update
	High	Low	Low	software product
	—»—	—»—	Mean	
	—»—	—»—	High	
	—»—	Mean	Low	
	—»—	—»—	Mean	
	Mean	High	—»—	
	High	—»—	Low	
	—»—	—»—	Mean	
	—»—	—»—	High	

The developed recommendations were applied to software products with rating points provided in Table 6.8.

According to these recommendations the software products No. 3, 4, 5, 6 and 8 were transferred to the next development cycle. For other products distances from their values of characteristics to values of corresponding characteristics of the reference pattern were determined, and these data were summarized in Table 6.13.

According to the developed managerial instructions the software products No. 7 and No. 11 were conditionally transferred to the next stage, other software products were offered to be updated by a group of experts.

References

1. Borisov, L.N., Alexeyev, A.V., Merkureva, G.V., et al.: Processing of the fuzzy information in decision-making systems. Radio i svyaz, Moscow (1989)
2. Pfanntzgal, I.: Theory of measuring: Transl. from English. Mir, Moscow (1976)
3. Kolmogorov, A.N.: Basic concepts of probability theory, 2nd edn. Nauka, Moscow (1974)
4. Zadeh, L.A.: Concept of a linguistic variable and its application to acceptance of approximate solutions. Mir, Moscow (1976)
5. Ayvazyan, S.A., Yenyukov, I.S., Meshalkin, L.D.: The applied statistics. Fundamentals of simulation and data preprocessing. Finansy i statistika, Moscow (1983)
6. Ayvazyan, S.A., Yenyukov, I.S., Meshalkin, L.D.: The applied statistics. Researches of Relations. Finansy i statistika, Moscow (1985)
7. Ayvazyan, S.A., Yenyukov, I.S., Meshalkin, L.D.: The applied statistics. Classification and dimension lowering. Finansy i statistika, Moscow (1989)
8. Nalimov, V.V., Chernova, N.L.: Statistical methods of planning of extreme experiments. Fizmatgiz, Moscow (1965)
9. Nalimov, V.V.: Mathematical theory of experiment. MEPhI, Moscow (1982)
10. The applied statistics. Data processing methods. The basic requirements and characteristics, 64 p. All-Union RSI of Standardization, Moscow (1987)
11. Dreyper, N., Smith, G.: The applied regression analysis. Finansy i statistika, Moscow (1987)
12. Shiryaev, A.N.: Probability. Nauka, Moscow (1980)
13. Tyurin, Y.N., Litvak, B.G., Orlov, A.I., et al.: The analysis of the non-numerical information, 80 p. Research council of Academy of Sciences of the USSR on a complex problem "Cybernetics", Moscow (1981)
14. Zadeh, L.A.: Fuzzy sets. Inform. and Control 8, 338–352 (1965)
15. Averkin, A.N., Batyrshin, I.Z., Blishun, A.F., et al.: Fuzzy sets in models of control and artificial intelligence. Main office on physical-math. literature, 312 p. Nauka, Moscow (1986)
16. Litvak, B.G.: Expert information: deriving and analysis methods, 184 p. Radio i svyaz, Moscow (1982)
17. Smolyak, S.A., Titarenko, B.P.: Steady evaluation methods: Statistical processing of inhomogeneous populations, 208 p. Statistika, Moscow (1980)
18. Genie, K.: Average quantities, 556 p. Statistika, Moscow (1970)
19. Orlov, A.I.: Problems of optimisation and fuzzy variables, 64 p. Znanie, Moscow (1980)
20. Vorobyov, O.Y., Valendik, E.N.: Probabilistic set simulation of distribution of forest fires, 160 p. Nauka, Novosibirsk (1978)
21. Yegorshin, A.: Forecast (About educational prospects in Russia). Higher Education in Russia 4, 17–21 (2000)
22. Alchinov, V., Kuptsov, A.: Rating-control of progress of cadets. Higher Education in Russia 1, 95–97 (1998)
23. Aseev, N., Dudkina, N., Fedorov, A.: Evaluation of a teacher's skill. Higher education in Russia 3, 41–46 (2001)
24. Zhurakovskiy, V., Prikhodko, V., Fedorov, I.: Today and tomorrow of an academician teacher – (Pedagogical and qualification aspects). Higher education in Russia 3, 3–12 (2000)

25. Martynova, T., Nadeljasv's, v., et al.: Rating system of knowledge evaluation at study of general engineering disciplines. Higher education in Russia 2, 103–107 (1997)
26. Melnichuk, O., Yakovleva, A.: Model of a specialist (To a problem about humanization of education). Higher education in Russia 5, 19–25 (2000)
27. Panin, M.: Morphology of a rating. Higher education in Russia (4), 90–94 (1998)
28. Ryzhov, A.P.: Theory of fuzzy sets and fuzziness measurement elements, 116 p. Dialog-MGU, Moscow (1998)
29. Khampel, F., Ronchetti, E., Rausseu, P., Shtael, V.: Robastnost in the statistics. The approach on the basis of influence functions, 512 p. Mir, Moscow (1989)
30. Candall, M., Stewarth, A.: Statistics conclusions and connections, 899 p. Nauka, Moscow (1973)
31. Hughber, P.: Robastnost in the statistics, 304 p. Mir, Moscow (1984)
32. Orlov, A.I.: Stability in social and economic models, 296 p. Nauka, Moscow (1979)
33. Litvak, B.G.: Expert evaluations and a decision making, 271 p. The patent, Moscow (1996)
34. Beshelev, S.D., Gurvich, F.G.: Matematical-statistics methods of expert evaluations, 2nd edn., 263 p. Statistika, Moscow (1980)
35. Malyshev, N.G., Bershteyn, L.S., Bozhenyuk, A.V.: Fuzzy models for expert systems in CAD systems, p. 136. Energoatomizdat, Moscow (1991)
36. Strength calculation norms of the equipment and pipelines of atomic power installations PNAE G-7 002-86. Energoatomizdat, Moscow (1989)
37. Nedosekin, A.O., Maksimov, O.B.: The analysis of bankruptcy risk of an enter-prise with application of fuzzy sets. Problems of the analysis of risk (2-3) (1999)
38. Ryzhov, A.P., A.N.: Axiomatic definition of degree of fuzziness of a linguistic scale and its basic characteristics. In: II All-Union conference - Artificial intellect 1990, Sectional and poster reports, Minsk, vol. 1, pp. 162–165 (1990)
39. Ryzhov, A.P.: Degree of fuzziness of a linguistic scale and its characteristics. In: Averkin, A.N., et al. (eds.) Fuzzy systems of support of a decision making, pp. 82–92. Publishing house of Kalinin State University, Kalinin (1988)
40. Kofman, A.: Introduction in the theory of fuzzy sets, 432p. Radio i svyaz, Moscow (1982)
41. Dubois, D., Prade, H.: Fuzzy Sets and Systems: Theory and Application, 393 p. Acad. press, N. Y (1980)
42. Dubois, D., Prade, H.: Ranking Fuzzy Numbers in Setting of Possibility Theory. Information Science 30, 183–224 (1983)
43. Dubois, D., Prade, H.: Fuzzy real algebra: some results. Fuzzy Sets and Systems 2(4), 327–348 (1979)
44. Mizumolo, M., Tanaka, K.: Algebraic characteristics of fuzzy numbers. In: Proc. IEEE Int. Conf. Cybernetics and Society, pp. 559–563 (1976)
45. Brutyan, G.A.: Hypothesis of Sepir-Uorf. Yerevan, 120p (1968)
46. Pospelov, D.A. (ed.): Models of a decision making on the basis of a linguistic variable, 312p. Nauka, Moscow (1986)
47. Borisov, A.N., Alexeev, A.V., Krumberg, O.A., et al.: Models of a decision making on the basis of a linguistic variable, 256 p. Zinatne, Riga (1982)
48. Averkin, A.N.: Fuzzy sets in artificial intellect models. Cybernetics problems. Artificial intellect problems (61), 79–86 (1980)
49. Borisov, A.N., Krumberg, O.A., Fedorov, I.P.: Decision making on the basis of fuzzy models: Examples of use, 184 p. Zinatne, Riga (1990)

50. Ryzhov, A.P.: An evaluation of degree of fuzziness and its application in artificial intellect systems. Intellectual systems 1(1-4), 95–102 (1996)
51. Ryjov, A.: Fuzzy data bases: description of objects and retrieval of information. In: Proceeding of the First European Congress in Intelligent Technologies, Aachen, Germany, vol. 3, pp. 1557–1562 (1993)
52. Forsyte, R. (ed.): Expert systems. Principles of work and examples: Transl. from English, p. 224. Radio I svyaz, Moscow (1987)
53. Law, C.-K.: Using fuzzy numbers in educational grading system. Fuzzy Sets and Systems 83, 311–323 (1996)
54. Nedosekin, A.O., Maksimov, O.B.: The analysis of risk of investments with application of fuzzy sets. Control of Risk 1 (2001)
55. Altunin, A.E., Semukhin, M.V.: Models and algorithms of a decision making in fuzzy conditions, 268 p. Publishing house of Tyumen State University, Tyumen (2002)
56. Alexeev, A.V.: Interpretation and definition of membership functions of fuzzy sets. In: Methods and systems of a decision making, pp. 42–50. RPI, Riga (1979)
57. Ragade, R.K., Gupta, M.M.: Fuzay sets theory, introduction. In: Gupta, M.M., Saridis, G., Gaines, B. (eds.) Fuzzy Automata and Decision Processes, pp. 105–131. North-Holland, Amsterdam (1977)
58. Thole, U., Zimmermann, H.J., Zysno, P.: On the suitability of minimum and products operators for the intersection of fuzzy sets. Fuzzy Sets and Systems 2, 167–180 (1979)
59. Zadeh, L.A.: Calculus of fuzzy restrictions. In: Zadeh, L.A., et al. (eds.) Fuzzy Sets and Their Applications to Cognitive and Decision Processes, pp. 1–41. Academic Press, New York (1975)
60. Alexeev, A.V.: Development of principles of application of the theory of fuzzy sets in situational models of control of organizational systems. Synopsis of a thesis. Cand. Sci. Eng., p. 20. RPI, Riga (1979)
61. Borisov, A.N., Ya, O.Y.: A technique of an evaluation of membership func-tions of elements of diffusion set. In: Cybernetics and Diagnostics, pp. 125–134. RPI, Riga (1970)
62. Loginov, V.I.: About probability treatment of Zadeh membership functions and their application for pattern recognition. Bulletins of AS of the USSR: Engineering cybernetics 2, 72–73 (1966)
63. Ya, O.Y.: Recognition of inaccuracies of complex units with use of fuzzy sets. In: Cybernetics and Diagnostics, pp. 13–18. RPI, Riga (1968)
64. Gridina, E.G., Lebedev, A.N.: A new method of definition of membership functions of fuzzy sets. New informational process engineerings 7, 30–33 (1997)
65. Scala, H.J.: On many-valued logics, fuzzy sets, fuzzy logics and their applications. Fuzzy Sets and Systems 1, 129–149 (1978)
66. Zhukovin, V.E., Oganesyan, N.A., Burshtein, F.V., Korelov, E.S.: About an approach to problems of a decision making from a position of the theory of fuzzy sets. In: Methods of a decision making in the conditions of uncertainty, pp. 12–16. RPI, Riga (1980)
67. Saaty, T.L.: Exploring the interface between hierarchies, multiple objectives and fuzzy sets. Fuzzy Sets and Systems 1, 57–69 (1978)
68. Rotshteyn, A.P., Shtovba, S.D.: The fuzzy multicriteria analysis of variants with application of paired comparisons. Bulletins of Academy of Sciences The theory of control and control systems 3, 150–154 (2001)

69. Blishun, A.F.: Simulation of decision-making process in fuzzy conditions on the basis of similarity of concepts of classes. Synopsis of a thesis. Cand. Sci. Phys. and Mathem., 19 p. VTs of AS of the USSR, Moscow (1982)

70. Saaty, T.L.: Measuring the fuzziness of sets. Journal of Cybernetics 4, 53–61 (1974)

71. Kikvidze, Z.A., Tkemaladze, N.T.: About one mode of weighing of elements of fuzzy set. Reports of AS of the USSR 93(2), 317–320 (1979)

72. Sher, A.P.: Coordination of fuzzy expert evaluations and membership function in a method of diffusion sets. In: Simulation and research of automatic control systems, pp. 111–118. DVNTs of AS of the USSR, Vladivostok (1978)

73. Ezhkova, I.V.: Semantically invariant formalisation of linguistic evaluations. In: Semiotics aspects of formalisation of intellectual activity, pp. 48–51. MDNTP, Moscow (1983)

74. Ezhkova, I.V., Pospelov, L.A.: Decision making at the fuzzy founda-tions. Bulletins of AS of the USSR: Engineering cybernetics 6, 3–11 (1977)

75. Svarovskiy, S.G.: Approximation of membership functions of values of a lin-guistic variable. In: Mathematical problems of the data analysis, pp. 127–131. NETI, Novosibirsk (1980)

76. Hodashinsky, I.A.: Logics of value evaluations. In: VINITI, Tomsk, vol. 1631-84, 19 p (1984)

77. Skofenko, A.V.: About building of membership functions of the fuzzy sets corresponding to quantitative expert evaluations. In: Research on research and computer science, vol. 22, pp. 70–79. Naukova dumka, Kiev (1981)

78. Borisov, A.N., Fomin, S.: The axiomatic approach to restore of membership function of terms of a linguistic variable. In: Model of a choice of alternatives in an fuzzy medium, pp. 77–79. RPI, Riga (1980)

79. Averkin, A.N.: Building of fuzzy models of the world for planning in the conditions of uncertainty. In: Semiotics models in control of big systems, pp. 69–73. AS of the USSR, Moscow (1979)

80. Zadeh, L.A.: A theory of approximate reasoning (AR). Machine Intelligence 9, 149–194 (1979)

81. Zadeh, L.A.: Fuzzy logic and approximate reasoning. Synthese 80, 407–428 (1975)

82. Kruglov, V.V., Dli, M.I., Yu, G.R.: The fuzzy logic and artificial neural webs, 224 p. Fizmatli, Moscow (2001)

83. Zadeh, L.A.: A basis of the new approach to the analysis of complex systems and decision-making processes. In: Moiseyev, N.N. (ed.) Mathematics today, pp. 5–48. Znanie, Moscow (1974)

84. Chang, Y.-H.: Hybrid fuzzy least-squares regression analysis and its reliability measures. Fuzzy Sets and Systems 119, 225–246 (2001)

85. Shkondin, A.I.: A fuzzy conclusion on the basis of definiteness modifiers. Bulletins of Academy of Sciences. The Theory of Control and Control Systems 1, 136–142 (2001)

86. Nasibov, E.N.: About two average characteristics of fuzzy numbers. VINITI 91(772), 24 (1991)

87. Nasibov, E.N.: Some integral indexes of fuzzy numbers and visual interactive method of definition of strategy of their evaluation. Bulletins of Academy of sciences. The Theory of Control and Control Systems 4, 82–88 (2002)

88. Sher, A.P.: Solution of a problem of mathematical programming with linear function in diffusion restrictions. Automation and Telemechanics 7, 17–22 (1980)

89. Llena, J.: On fuzzy linear programming. EJOR 22(2), 124–131 (1985)

90. Yager, R. (ed.): Fuzzy sets and the theory of possibilities, p. 408. Radio I svyaz, Moscow (1986)
91. Nguyen, H.T.: A note on the extension principle for fuzzy sets. Journal of Mathematical Analysis and Application 64, 369–380 (1978)
92. Mizumoto, M., Zimmerman, H.J.: Comparison of fuzzy reasoning methods. Fuzzy Sets and Systems 8, 253–283 (1982)
93. Domrachev, V.G., Poleshchuk, O.M.: About building of regression models at fuzzy input data. Automation and Telemechanics 11, 74–83 (2003)
94. Zhuravlyov, Y.I.: About the algebraic approach to a solution of problems of recognition and classification. Cybernetics Problems 33, 26–57 (1978)
95. Kudryavtsev, V.B., Alyoshin, S.T., Podkolzin, A.S.: Introduction in the theory of automatic machines, 319 p. Nauka, Moscow (1985)
96. Drobyshev, Y. P., Pukhov, V.V.: Approximatin of fuzzy relations. In: Empirical prediction and pattern recognition (Computing systems), Novosibirsk, vol. 76, pp. 75–82 (1978)
97. Mirkin, B.G.: Analysis of qualitative indexes and structures, 319 p. Statistika, Moscow (1982)
98. Zadeh, L.A.: Fuzzy sets and their application in a pattern recognition and the cluster analysis. In: Van Raisin, J. (ed.) Classification and a cluster, pp. 208–247. Mir, Moscow (1980)
99. Ruspini, E.G.: Latest advances in a fuzzy cluster analysis/. In: Yager, R.R. (ed.) Fuzzy Sets and The Theory of Possibilities The latest advances, pp. 47–62. Radio i svyaz, Moscow (1986)
100. Tamura, S., Higuchi, S., Tanaka, K.: Pattern classification based on fuzzy relalions. IEEE Transactions on Systems, Man and Cybernetics SMC-1, 61–66 (1971)
101. Zadeh, L.A.: Similarity relations and fuzzy orderings. Information Sciences 3, 177–200 (1971)
102. Ruspini, E.H.: A new approach to clustering. Information and Control 15, 22–32 (1969)
103. Ruspini, E.H.: Numerical methods for fuzzy clustering. Information Sciences 2, 319–350 (1970)
104. Batyrshin, I.Z.: Clusterization on the basis of diffusion relations of similarity. In: Control in the presence of fuzzy categories: Theses of reports of III scientific and technical seminar, pp. 25–27. NIIUMS, Perm (1980)
105. Duda, R., Heart, P.: Pattern recognition and the analysis of scenes. Mir, Moscow (1976)
106. Duran, B., Odell, P.: Cluster analysis, 128 p. Statistics, Moscow (1977)
107. Yeliseyev, I.I., Rukavishnikov, P.O.: Grouping, correlation, pattern recognition, 144 p. Statistika, Moscow (1977)
108. Jacque-Lagrez, E.: Application of diffusion relations at an evaluation of preference of the distributed values. In: Statistical models and multicriteria problems of a decision making, pp. 168–183. Statistika, Moscow (1979)
109. Kononov, B.P.: Numerical relations in the theory of comparisons. Scientific and technical information (NTI) 2, 11–18 (1977)
110. Kuzmin, V.B.: Building of group solutions in spaces of definite and fuzzy binary relations, 168 p. Nauka, Moscow (1982)
111. Bezdek, J.C.: Numerical taxonomy with fuzzy sets. Journal of Mathematical Biology 1, 57–71 (1974)

112. Negoita, C.V., Ralescu, D.A.: Applications of fuzzy sets to systems analysis, 187 p. Birkauser Verlag, Basel (1975)

113. Kaufmann, A.: Introduction to the theory of fuzzy subsets, vol. 1, 643 p. Academic Press, New York (1975)

114. Rosenfeld, A.: Fuzzy graphs. In: Zadeh, L.A., et al. (eds.) Fuzzy Sets and Their Applications to Cognitive and Decision Processes, pp. 77–95. Academic Press, New York (1975)

115. Ruspini, E.H.: Recent developments in fuzzy clustering. In: Yager, R.R. (ed.) Fuzzy Set and Possibility Theory, pp. 133–146. Pergamon Press, New York (1982)

116. Shimura, M.: Fuzzy sets concepts in rank-ordering objects. Journal of Mathematical Analysis and Applications 43, 713–733 (1973)

117. Watada, J., Tanaka, H., Asai, K.: A heuristic method of hierarchical clustering for fuzzy intransitive relations. In: Yager, R.R. (ed.) Fuzzy Set and Possibility Theory, pp. 148–166. Pergamon Press, New York (1982)

118. Yeh, R.T., Bang, S.Y.: Fuzzy relations, fuzzy graphs and their applications to clustering analysis. In: Zadeh, L.A., et al. (eds.) Fuzzy Sets and Their Applications to Cognitive and Decision Processes, pp. 125–149. Academic Press, New York (1975)

119. Zadeh, L.A.: Fuzzy sets and their application to pattern classification and cluster analysis. Memo No. ERL M 607. Electronics Research Laboratory. California University, Berkeley, 67 p (1976)

120. Negoita, C.V., Ralescu, D.A.: Representation theorems for fuzzy concepts. Kybernetes 4, 169–174 (1975)

121. Poleshchuk, O.M.: Metods of representation of the expert information in the form of a population of term-sets of complete orthogonal semantic spaces. Bulletin of Moscow State Forestry University — Forestry Bulletin 5(25), 198–216 (2002)

122. Poleshchuk, O.M.: About development of systems of processing of fuzzy information on the basis of complete orthogonal semantic spaces. Bulletin of Moscow State Forestry University — Forestry Bulletin 1(26), 112–117 (2003)

123. Saati, T.: A method of the analysis of hierarchies, 320 p. Radio i svyaz, Moscow (1993)

124. Levin, V.I.: New generalisation of operations over fuzzy sets. Bulletins of Academy of sciences. The Theory of Control and Control Systems 1, 143–146 (2001)

125. Poleshchuk, O.M.: Application of semantic spaces for an expert evaluation of characteristics of quality of software and a fuzzy multicriteria choice. Bulletin of Moscow State Forestry University — Forestry Bulletin 1(32), 120–125 (2004)

126. Domrachev, V.G., Poleshchuk, O.M., Popova, I.A.: Model of multicriterial expert evaluation of software for providing financial and economic activity of an organisation. Ibidem 4(35), 149–153

127. Domrachev, V.G., Poleshchuk, O.M., Retinsky, I.V.: Selection of software on the basis of characteristics of quality and system of fuzzy production rules. In: First scientific conference "Quality and IPI (CALS) – technologies", Conference materials, Sudak, pp. 53–54 (2004)

128. The general technique of an evaluation of software quality, Moscow, p. 53 (1988)

129. Hubaev, G.N.: An economic evaluation of want-satisfying qualities of software: Texts of lectures, RostovonDon. RGEA, 104 p. (1997)

130. Lipaev, V.V.: Software quality ensuring. Methods and standards, 355 p. STU "Stankin", Moscow (2000)

131. Criteria for Evaluation of Software. ISO TC97/SC7 # 383

132. Revised version of DP9126 — Criteria of the Evaluation of Software Quality Characteristics. ISO TC97/SC7 #610. — Part 6

133. Boem, B., Brown, J., Kaspar, K., et al.: Characteristics of quality of program security, pp. 61–87. Mir, Moscow (1981)

134. Poleshchuk, O.M.: About evaluation of analogousness of engineering products within the limits of some parameter. Review of Applied and Industrial Mathematics 10(1), 205 (2003)

135. Poleshchuk, O.M., Poleshchuk, I.A.: Fuzzy clusterization of elements of set of complete orthogonal semantic spaces. Bulletin of Moscow State Forestry University — Forestry Bulletin 1(26), 117–127 (2003)

136. Poleshchuk, O.M.: Methods of preliminary processing of the fuzzy expert information at a stage of its formalisation. Ibid 5(30), 160–167

137. Hoffmann, O.G.: Expert evaluation, 152 p. Publishing House of VGU, Voronezh (1991)

138. Candall, M.: Ranking correlations. Transl. from English, 214 p. Statistika, Moscow (1975)

139. Vinnkov, B.G., Gokhman, A.O., Gokhman, O.G.: Evaluation of a consistency of expert judgements by preparation of the information for curriculum calculation on a speciality: Cybernetics methods and means in control of educational process of the higher school, Riga, pp. 55–61 (1987)

140. Mirkin, B.G.: Grouping in social and economic researches: Building and analysis methods, 222 p. Finansy i statistika, Moscow (1985)

141. Arrow, K.J.: Social Choice and Individual values, 124 p. Yale Univ. Press, New Haven (1972)

142. Milenkiy, A.B.: Classification of signals in conditions the indefinity, 328 p. Sov. radio, Moscow (1975)

143. Kemeni, J., Snell, J.: Cybernetic model, 192 p. Sov. radio, Moscow (1972)

144. Lezina, Z.M.: Processes of a collective choice. Automation and Telemechanics 8, 3–35 (1987)

145. Hwang, C.L., Lin, N.J.: Group decision making under multiple criteria, 400 p. Springer, Berlin (1987)

146. Komarov, E.G., Poleshchuk, I.A., Poyarkov, N.G.: Models of processing of the information in educational process on the basis of methods of the theory of fuzzy sets. Telecommunications and Informational Support of Education 1, 69–80 (2006)

147. Poleshchuk, O.M.: Mathematics model for processing of expert evaluations. Bulletin of Moscow State Forestry University — Forestry Bulletin 6(42), 161–164 (2005)

148. Mozolevskaya, E.G.: Monitoring of a condition of plantings and city woods of Moscow. Methods of an evaluation of a condition of trees and plantings. In: Ecology of a big city, vol. 2, pp. 16–59. Prima-Press, Moscow (1997)

149. Frolov, V.A.: About a condition of green plantings in territory of parkways of the southwest of Moscow (by results of monitoring of 1998). In: Problems of quality management of environment: Collection of Reports of International Confer., pp. 202–204. Prima-Press, Moscow (1999)

150. Poleshchuk, O.M.: Building of integral models in frameworks the fuzzy expert information. Bulletin of Moscow State Forestry University — Forestry Bulletin 5(30), 155–159 (2003)

151. Fishburn, P.: Theory of utility for a decision making. Nauka, Moscow (1978)

152. Coleman, T.F., Li, Y.: A reflective newton method for minimising a quadratic function subject to bounds on some of the variables. SIAM J. Optim. 6(4), 1040–1058 (1996)

153. Dubois, D.: Linear Programming with fuzzy data. Analysis of Fuzzy Information 3, 241–263 (1987)

154. Poleshchuk, O., Komarov, E.: The determination of rating points of objects with qualitative characteristics and their usage in decision making problems. Proceedings of World Academy of Science, Engineering and Technology 40, 313–317 (2009), ISSN: 2070-3740

155. Poleshchuk, O., Komarov, E.: The determination of rating points of objects and groups of objects with qualitative characteristics. In: Proceedings of the 28th International Conference of the North American Fuzzy Information Processing Society, NAFIPS 2009, Cincinnati, Ohio, June 14-17 (2009), ISBN: 978-1-4244-4577-6

156. Poleshchuk, O., Komarov, E.: The determination of students' fuzzy rating points and qualification levels. In: Proceedings of the 1st International Fuzzy Systems Symposium- FUZZYSS 2009, Ankara, Turkey, pp. 218–224 (2009)

157. Poleshuk, O.M., Komarov, E.G.: The determination of students' rating points on fuzzy formalization of initial information basis. In: Education, science and economics at universities, Integration to international education area, Plock, Poland, pp. 67–73 (2008)

158. Poleshuk, O.M., Komarov, E.G.: The determination of the rating points of the students' groups. In: Education, science and economics at universities. Integration to international education area, Plock, Poland, pp. 74–79 (2008)

159. Cheng, C.H., Wang, J.W., Tsai, M.F., Huang, K.C.: Appraisal support system for high school teachers based on fuzzy linguistic integrating operation. Journal of Human Resource Management (4), 73–89 (2004)

160. Wang, C.H., Chen, S.M.: Appraising the performance of high school teachers based on fuzzy number arithmetic operations. Soft Computing -A Fusion of Foundations, Methodologies and Applications (12), 919–934 (2008)

161. Wang, Y.H., Yang, J.B., Xu, D.L., Chin, K.S.: On the centroids of fuzzy Numbers. Fuzzy Sets and Systems 157, 919–926 (2006)

162. Echauz, J.R., Vachtsevanos, G.J.: Fuzzy grading system. IEEE Trans. Educ. 38(2), 158–164 (1995)

163. Ranjit, B.: An application of fuzzy set in students' evaluation. Fuzzy Sets and Systems 74, 187–194 (1995)

164. Biswas, R.: An application of fuzzy sets in student's evaluation. Fuzzy Set and systems 74, 194–197 (1995)

165. Capaldo, G., Zollo, G.: Applying fuzzy logic to personnel assessment: A case study. Omega The International Journal (29), 585–597 (2001)

166. Chen, S.M., Lee, C.H.: New methods for students' evaluation using fuzzy sets. Fuzzy Sets and Systems 104, 209–218 (1999)

167. Hubka, V.: Theory of engineering systems, 208 p. Mir, Moscow (1987)

168. Khubyaev, G.: About building of a rating scale and testing systems. Higher Education in Russia 1, 122–125 (1996)

169. Ershikov, S., Lobov, T., Phillipov, S., Shidlovska, T.: Experience of use of rating system. Higher Education in Russia 4, 97–102 (1997)

170. Kruglikov, V.: Rating system of diagnostics of educational process of a high school. Higher education in Russia 2, 100–102 (1996)

171. Litvak, B.G.: Expert evaluations and a decision making, 271 p. Patent, Moscow (1996)
172. Poleshchuk, O.M.: Building of ratings with use of linguistic variables. In: Materials of the All-Russia Scientific Conference, Saransk, pp. 104–108 (2002)
173. Poleshchuk, O.M.: About building of ratings on a basis of linguistic variables. Bulletin of Moscow State Forestry University — Forestry Bulletin 3(28), 169–177 (2003)
174. Poleshchuk, O.M., Retinskaya, I.V.: Building of ranking scores with use of complete orthogonal semantic spaces. In: Provorova, O.G. (ed.) II All-Russia congress of women-mathematicians, p. 160. Krasnoyarsk State University (2002)
175. Podinovskiy, V.V.: Quantitative evaluations of importance of criteria of multicriteria optimization. NTI. Ser.2 - VINITI 5, 22–25 (1999)
176. Poleshchuk, O.M.: About application of fuzzy sets in problems of building of level gradation. Forestry Bulletin 4(13), 142–146 (2000)
177. Poleshchuk, O.M., Retinskaya, I.V.: Problem of personnel selection and linguistic variables. In: Provorova, O.G. (ed.) II All-Russia congress of women - mathematicians, p. 162. Krasnoyarsk State University (2002)
178. Domrachev, V.G., Petrov, V.A., Poleshchuk, O.M.: Linguistic variables in problems of personnel selection. Forestry Bulletin 5(20), 192–197 (2001)
179. Domrachev, V.G., Petrov, V.A., Poleshchuk, O.M.: About application of linguistic variables in problems of personnel selection. In: Informational and measuring technics, ecology and monitoring, vol. 1, pp. 554–560. Publishing House MGUL, Moscow (2001)
180. Poleshchuk, O.M.: Occupational selection of graduates on the basis of fuzzy characteristics. Bulletin of Moscow State Forestry University — Forestry bulletin 4(35), 142–149 (2004)
181. Poleshchuk, O.M.: About application of the theory of fuzzy sets means in problems of processing of the information in educational process. Ibid 3(28), 164–169 (2003)
182. Poleshchuk, O.M.: Some approaches to control system model in educational process. Telecommunications and Information Support in Education 3(10), 54–72 (2002)
183. Poleshchuk, O.M.: Application of linguistic variables in modeling of educational process. Forestry Bulletin 2(16), 194–195 (2001)
184. Baas, S.M., Kwakernaak, H.: Rating and ranking of multiple-aspect alternative using fuzzy sets. Automatica 3(1), 47–58 (1977)
185. Poleshchuk, O.M., Frolova, V.A.: Ranking scores of a condition of city plantings on the basis of methods of the theory of fuzzy sets. Forestry Journal 2-3, 7–14 (2003)
186. Komarov, E.G., Poleshchuk, O.M., Frolova, V.A.: About determination of ranking scores for a condition of species of the plants growing in complex ecological conditions of big cities. Review of the Applied and Industrial Mathematics 10(1), 175 (2003)
187. Nedosekin, A.O., Maximov, O.B.: Analysis of bankruptcy risk of an enterprise with application of fuzzy sets. Problems of the Analysis of Risk 2-3 (1999)
188. Chang, Y.-H., Ayyub, B.M.: Fuzzy regression methods — a comparative assessment. Fuzzy Sets and Systems 119, 187–203 (2001)
189. Poleshchuk, O., Komarov, E.: Hybrid fuzzy least-squares regression model for qualitative characteristics. In: Huynh, V.-N., Nakamori, Y., Lawry, J., Inuiguchi, M. (eds.) Integrated Uncertainty Management and Applications. AISC, vol. 68, pp. 187–196. Springer, Heidelberg (2010)
190. Tanaka, H., Asai, U.S.: Linear regression analysis with fuzzy model. IEEE. Systems, Trans. Systems Man Cybernet. SMC-2, 903–907 (1982)

191. Chanrong, V., Haimes, y.Y.: Multiobjective Decision Making: Theory and Methodology, p. 204. North-Holland, Amsterdam (1980)

192. Dubois, D.: Linear Programming with fuzzy data. Analysis of Fuzzy Information 3, 241–263 (1987)

193. Fiacco, A.V.: Introduction to sensitivity and stability analysis in nonlinear programming, 367 p. Acad. Press, New York (1983)

194. Tarantsev, A.A.: Building principles for regression models at initial fuzzy exposition data. Automation and Telemechanics 41, 27–32 (1997)

195. Tanaka, H.: Fuzzy data analysis by possibilistic linear models. Fuzzy Sets and Systems 24, 363–375 (1987)

196. Sakawa, M., Yano, H.: Fuzzy linear regression and its application to the sales borecasting. International Journal of Policy and Information 15, 111–125 (1989)

197. Sabic, D.A., Pedrycr, W.: Evaluation on fuzzy linear regression models. Fuzzy Sets and Systems 39, 51–63 (1991)

198. Tanaka, H., Watada, J.: Possibilistic linear systems and their applications to the linear regression model. Fuzzy Sets and Systems 27, 275–289 (1988)

199. Bardossy, A.: Note on fuzzy regression. Fuzzy Sets and Systems 37, 65–75 (1990)

200. Celmins, A.: Least squares model fitting to fuzzy vector data. Ibid 22, 245–269 (1987)

201. Celmins, A.: Multidimensional least-squares model fitting of fuzzy models. Math. Modeling 9, 669–690 (1987)

202. Chang, Y.-H., Ayyub, B.M.: Reliability analysis in fuzzy regression. In: Proc. Annual Conf. of the North America fuzzy Information Society (NAFTPS 1993), Allentown, pp. 93–97 (1993)

203. Chang, P.-T., Lee, E.S.: Fuzzy linear regression with spreads unrestricted in sign. Comput. Math. Appl. 28, 61–71 (1994)

204. Chang, Y.-H., Johnson, P., Tokar, S., Ayyub, B.M.: Least-squares in fuzzy regression. In: Proc. Annual. Conf. of die North America fuzzy Information Society (NAFTPS 1993), Allentown, pp. 98–102 (1993)

205. Diamond, P.: Fuzzy least squares. Inform. Set. 46, 141–157 (1988)

206. Ishibuchi, H.: Fuzzy regression analysis. Japan. J. Fuzzy Theory and Systems 4, 137–148 (1992)

207. Redden, D., Woodal, W.: Characteristics of certain fuzzy regression methods. Fuzzy Sets and Systems 64, 361–375 (1994)

208. Tanaka, H., Ishibuchi, H.: Identification of possibilistic linear models. Fuzzy Sets and Systems 41, 145–160 (1991)

209. Tanaka, H., Ishibuchi, H., Yoshikawa, S.: Exponential possibility regression analysis. Fuzzy Sets and Systems 69, 305–318 (1995)

210. Celmins, A.: A practical approach to nonlinear fuzzy regression. SIAM. J. Sci & Stat. Computing 12, 329–332 (1991)

211. Poleshchuk, I.A.: Multi-dimensional nonlinear regression model at fuzzy in-put data. Bulletin of Moscow State Forestry University — Forestry Bulletin 6(42), 164–172 (2005)

212. Domrachev, V.G., Poleshchuk, I.A.: About square regression models at fuzzy input data. Review of Applied and Industrial Mathematics 10(3), 647–648 (2003)

213. Moore, R.E.: Interval Analysis. Prentice-Hall, Englewood Cliffs (1966)

214. Moore, R.E.: Methods and Applications of Interval Analysis. SIAM, Philadelphia (1979)

215. Hathaway, R.J., Bezdek, J.C.: Switching regression models and fuzzy clustering. IEEE Transactions on Fuzzy Systems 1(3), 195–203 (1993)
216. Turksen, I.B.: Fuzzy functions with LSE. Applied Soft Computing 8(3), 1178–1188 (2008)
217. Celikyilmaz, A.: Fuzzy functions with support vector machines. M.A.Sc Thesis, Information Science, Industrial Engineering Department, University of Toronto (2005)
218. Celikyilmaz, A., Turksen, I.B.: Fuzzy functions with support vector machines. Information Sciences 177(23), 5163–5177 (2007)
219. Yao, C.C., Yu, P.T.: Fuzzy regression based on asymmetric support vector machines. Applied Mathematics and Computation 182, 175–193 (2006)
220. Shurygin, A.M.: Regression: a choice of a type of relation, efficiency and stability of evaluations. Automation and telemechanics 6, 90–101 (1996)
221. Poleshuk, O.M., Komarov, E.G.: New defuzzification method based on weighted intervals. In: Proceedings of the 27th International Conference of the North American Fuzzy Information Processing Society, NAFIPS 2008, New York, May 19-22 (2008)
222. Poleshuk, O.M., Komarov, E.G.: Multiple hybrid regression for fuzzy observed data. In: Proceedings of the 27th International Conference of the North American Fuzzy Information Processing Society, NAFIPS 2008, New York, May 19-22 (2008)
223. Domrachev, V.G., Poleshchuk, I.A.: Building of a reference at the comparative analysis of units with non-numerical characteristics. In: Records of Research and Practice Conference, KBD-Info 2004, Sochi, pp. 56–58 (2005)
224. Domrachev, V.G., Poleshchuk, I.A.: Building of ranking scores on the basis of a fuzzy conditional reference pattern. In: Records of Research and Practice Conference, IT - Innovations in education, Petrozavodsk, pp. 87–88 (2005)
225. Komarov, E.G., Poleshchuk, O.M., Poyarkov, N.G.: Definition of ranking score of entrants at initial fuzzy information. In: Records of Research and Practice Conference, KBD-Info-2005, Sochi, pp. 221–224 (2005)
226. Domrachev, V.G., Komarov, E.G., Poleshchuk, O.M., Poyarkov, N.G.: Application of methods of the fuzzy cluster analysis for improving of examination paper check quality. Ibid, 224–226
227. Komarov, E.G., Poyarkov, N.G.: Models of processing of the information of testing measures at a stage of pre-higher education training. Bulletin of Moscow State Forestry University — Forestry bulletin 1(43), 175–178 (2006)
228. Poleshchuk, I.A.: About connection between a fuzziness of the conditions superimposed on an information aggregation operator, and a fuzziness of a class of k-ary logic functions defined by these conditions. Ibid, 1061–1062
229. Poleshchuk, I.A.: About determination of an adequate operator of aggregation of the information. In: Records of Research and Practice Conference, KBD-Info-2004, Sochi, pp. 48–49 (2004)
230. Poleshchuk, I.A.: Selection of an adequate operator of aggregation of the information from a class of of k-ary logic functions. Bulletin of Moscow State Forestry University — Forestry bulletin 1(43), 170–174 (2006)

About the Authors

Olga Poleshchuk has graduated from the Moscow State University named after M. V. Lomonosov (Mechanical- Math. Faculty). In 1990 she earned a degree of PHD in Math. Sciences in Lomonosov Moscow State University, in 2004 she earned a degree of a Doctor of Tech. Sciences. Since 1993 she has been working in Moscow State Forest University. At the moment she is a professor. She has been a scientific tutor of three PHDes. She has got more than 100 scientific papers and a monograph.

Evgeniy Komarov has graduated from Moscow State Forest University (Electronics and Computers Faculty). In 1996 he earned a degree of PHD in Tech. Sciences. Since 1996 he has been working in Moscow State Forest University. At the moment he is a head of Department of Electronics and Microchips and a Vice-Rector. He has got more than 50 scientific papers, 9 article patents and a monograph.